Gene Amplification and Analysis

Volume 5
Restriction Endonucleases
and Methylases

Gene Amplification and Analysis
Jack G. Chirikjian and Takis S. Papas, *Series Editors*

VOLUMES ALREADY PUBLISHED

Volume 1—Restriction Endonucleases, Jack G. Chirikjian, ed.

Volume 2—Structural Analysis of Nucleic Acids, Jack G. Chirikjian and Takis S. Papas, eds.

Volume 3—Expression of Cloned Genes in Prokaryotic and Eukaryotic Cells, Takis S. Papas, Martin Rosenberg, and Jack G. Chirikjian, eds.

Volume 4—Oncogenes, Takis S. Papas and George F. Vande Woude, eds.

FORTHCOMING VOLUME IN THE SERIES

Volume 6—Application of Nucleic Acid Probe Technology in Medicine, Jack G. Chirikjian, ed.

Gene Amplification and Analysis

Volume 5
Restriction Endonucleases and Methylases

Edited by
Jack G. Chirikjian, Ph.D.
Department of Biochemistry
Vincent T. Lombardi Cancer Research Center
Georgetown University Medical Center
Washington D.C.

ELSEVIER
NEW YORK • AMSTERDAM • LONDON

Elsevier Science Publishing Co., Inc.
52 Vanderbilt Avenue, New York, New York 10017

Sole distributors outside the United States and Canada:
Elsevier Science Publishers, B.V.
P.O. Box 211, 1000 AE Amsterdam, the Netherlands

© 1987 by Elsevier Science Publishing Co., Inc.

This book has been registered with the Copyright Clearance Center, Inc.
For further information, please contact the Copyright Clearance Center, Inc.,
Salem, Massachusetts.

Library of Congress Cataloging in Publication Data

Restriction endonucleases and methylases.

    (Gene amplification and analysis, ISSN 0275-2778;
v.5)
    Includes index.
    1. Restriction enzymes, DNA. 2. Methyltransferases.
3. Molecular biology. I. Chirikjian, Jack G.
II. Series. [DNLM: 1. DNA Restriction Enzymes.
W1 GE 184N v.5/QU 136 R4362]
QP609.R44R48   1987   574.19'25   87-19915
ISBN 0-444-01285-0
ISSN 0275-2778

Current printing (last digit)
10 9 8 7 6 5 4 3 2 1

Manufactured in the United States of America

Contents

PREFACE   vii

CONTRIBUTORS   ix

Chapter 1

Restriction and Modification Enzymes and their Recognition Sequences   1
    Richard J. Roberts

Chapter 2

Restriction Endonuclease: Cleavage, Ligation, and Sensitivity   51
    Robert W. Blakesley

Chapter 3

Mechanism of Specific Site Location and DNA Cleavage by EcoR I
Endonuclease   103
    Brian J. Terry, William E. Jack, and Paul Modrich

Chapter 4

Structure and Function of the EcoR I Restriction Endonuclease   119
    John M. Rosenberg, Judith A. McClarin, Christin A. Frederick,
    John Grable, Herbert W. Boyer, and Patricia J. Greene

Chapter 5

The Enzymes of the BamH I Restriction-Modification System   147
    Glenn Nardone and Jack G. Chirikjian

Chapter 6

The EcoR V Restriction Endonuclease   185
    Peter A. Luke, Sarah A. McCallum, and Stephen E. Halford

Chapter 7

The Organization and Control of Expression of the *Pst* I Restriction-
Modification System   209
    Roxanne Y. Walder and Joseph A. Walder

Chapter 8

The *Pvu* II Restriction-Modification System: Cloning, Characterization and Use in Revealing an *E. coli* Barrier to Certain Methylases
or Methylated DNAs   227
    Robert M. Blumenthal

Chapter 9

Enzymatic Probes for Left-Handed Z-DNA   247
    Franz Wohlrab and Robert D. Wells

Chapter 10

Enhancement of the Apparent Cleavage Specificities of Restriction
Endonucleases: Applications to Megabase Mapping of Chromosomes   257
    Michael McClelland and Michael Nelson

Index   283

Preface

The mission of this series is to convey and update in a timely fashion significant information on selected aspects of molecular biology. The first four volumes of the series have concentrated on specific topics chosen for review. This volume, Restriction Endonucleases and Methylases, is the first to update an earlier volume in the series. Since the publication of Restriction Endonucleases Volume I in 1981, these enzymes have fulfilled their potential as powerful tools for molecular biology experiments. Over that same period of time, the number of reported enzymes have more than doubled, and several cognate restriction methylases have also found an important place in various experimental approaches. Of special note is the larger number of research groups that have focused their attention on studies that deal with the structure and catalytic properties of both restriction endonucleases and methylases.

The chapters in this volume represent various aspects of research in this area. The volume has several contributions that collectively serve as a useful reference for reaction conditions and the general application of restriction endonucleases as reagents for molecular biology. The rest of the volume reports studies that deal with the structure and enzymology of this important group of DNA enzymes.

I wish to take this opportunity to express my gratitude to the authors who are making contributions for the second time and those who joined our effort for the first time in making this volume possible. Special appreciation is also extended to Jill George for her excellent effort in organizing and typing the manuscripts.

Jack G. Chirikjian, Ph.D.
Georgetown University Medical Center
January, 1987

Contributors

Numbers in brackets indicate the page on which the author's contribution begins.

R.W. Blakesley, Molecular Biology R & D, Bethesda Research Laboratories, Life Technologies, Inc. 8717 Grovemont Circle, Gaithersburg, MD 20877 [50]

R. M. Blumenthal, Department of Microbiology, Medical College of Ohio, Toledo, OH 43699 [224]

H. W. Boyer, Department of Biology, Massachusetts Institute of Technology, 77 Massachusetts Avenue, Cambridge, MA 02139 [118]

J. G. Chirikjian, Department of Biochemistry, Georgetown University, Lombardi Cancer Center, 3800 Reservoir Road, Washington, DC 20007 [145]

C. A. Frederick, Department of Biological Sciences, University of Pittsburgh, Pittsburgh, PA 15260 [118]

J. Grable, Department of Biological Sciences, University of Pittsburgh, Pittsburgh, PA 15260 [118]

P. J. Greene, Department of Biochemistry and Biophysics, University of Calfiornia San Francisco, San Francisco, CA 94143 [118]

Stephen E. Halford, Department of Biochemistry, Unit of Molecular Genetics, University of Bristol, Bristol BS8 1TD (U.K.) [183]

W. E. Jack, Department of Biochemistry, Duke University Medical Center, Durham, NC 27710 [102]

P. A. Luke, Anglian Biotechnology Ltd., Hawkins Road, Colchester COZ 8JX (U.K.) [183]

S. A. McCallum, Department of Biochemistry, Unit of Molecular Genetics, University of Bristol, Bristol BS8 1TD (U.K.) [183]

J. A. McClarin, Department of Biological Sciences, University of Pittsburgh, Pittsburgh, PA 15260 [118]

M. McClelland, Department of Biochemistry and Molecular Biology, University of Chicago, 920 East 58th Street, Chicago, IL 60637 [251]

P. Modrich, Department of Biochemistry, Duke University Medical Center, Durham, NC 27710 [102]

G. Nardone, Department of Biochemistry, Georgetown University, Lombardi Cancer Research Center, 3800 Reservoir Road, Washington, DC 20007 [145]

M. Nelson, New England Biolabs, Inc., 32 Tozer Road, Beverly, MA 01915 [251]

R. J. Roberts, Cold Spring Harbor Laboratory, P.O. Box 100, Cold Spring Harbor, NY 11724 [1]

J. M. Rosenberg, Department of Biological Sciences, University of Pittsburgh, Pittsburgh, PA 15260 [118]

B. J. Terry, Department of Biochemistry, Duke University Medical Center, Durham, NC 27710 [102]

J. A. Walder, Department of Biochemistry, University of Iowa, Iowa City, IA 52242 [206]

R. Y. Walder, Department of Biochemistry, University of Iowa, Iowa City, IA 52242 [206]

R. D. Wells, Department of Biochemistry, Schools of Medicine and Dentistry, University of Alabama at Birmingham, Birmingham, AL 35294 [242]

F. Wohlrab, Department of Biochemistry, Schools of Medicine and Dentistry, University of Alabama at Birmingham, Birmingham, AL 35294 [242]

Gene Amplification and Analysis

Volume 5
Restriction Endonucleases
and Methylases

1

Restriction and Modification Enzymes and their Recognition Sequences

Richard J. Roberts

Cold Spring Harbor Laboratory
PO Box 100
Cold Spring Harbor, NY 11724

INTRODUCTION

Since the last published compilation of restriction endonucleases [234] 130 new enzymes have been discovered, including many new specificities. Especially noteworthy are the enzymes *Not*I (GCGGCCGC) and *Sfi*I (GGCCNNNNNGGCC) which are the first Type II enzymes to recognize octanucleotide sequences. They have the useful property of cutting DNA sufficiently infrequently so that their sites provide useful landmarks for mapping bacterial genomes and eukaryotic chromosomes. Among the 645 enzymes listed there are now a minimum of 137 different specificities.

In forming this list all endonucleases cleaving DNA at a specific sequence have been considered to be restriction enzymes, although in most cases there is no direct genetic evidence for the presence of a restriction-modification system. These endonucleases are named in accordance with the proposal of Smith and Nathans [277]. Within the table the source of each microorganism is given either as an individual or a national culture collection. If further information is required it can be found either in the first reference which in each case refers to the purification procedure for the restriction enzyme, or from the individuals who have provided their unpublished results. Where more than one reference appears, the second concerns the recognition sequence for the restriction enzyme, the third describes the purification procedure for the methylase and the fourth describes the recognition sequence of the methylase. In some cases, several references appear in one of these categories when independent groups have reached similar conclusions.

This work was supported by a grant from the National Science Foundation (PCM-8302902

Microorganism	Source	Enzyme	Sequence	λ	Ad2	SV40	ΦX	pBR	References
Acetobacter aceti	IFO 3281	AatI (StuI)	AGGCCT	6	11	7	1	0	289
		AatII	GACGT↑C	10	3	0	1	1	289
Acetobacter aceti sub. liquefaciens	IFO 12388	AacI (BamHI)	GGATCC	5	3	1	0	1	262
Acetobacter aceti sub. liquefaciens	M. Van Montagu	AaeI (BamHI)	GGATCC	5	3	1	0	1	262
Acetobacter aceti sub. orleanensis	NCIB 8622	AorI (BstNI)	CC↑(A/T)GG	71	136	17	2	6	262
Acetobacter liquefaciens	IAM 1834	AliI (BamHI)	G↑GATCC	5	3	1	0	1	335
Acetobacter liquefaciens AJ 2881	K. Yamada	AliAJI (PstI)	CTGCA↑G	28	30	2	1	1	246
Acetobacter pasteurianus	IFO 13753	ApaLI	G↑TGCAC	4	7	0	1	3	346
Acetobacter pasteurianus sub. pasteurianus	NCIB 7215	ApaI	GGGCC↑C	1	12	1	0	0	261
Acetobacter xylinus	IFO 3288	AxyI (SauI)	CC↑TNAGG	2	7	0	0	0	339
Achromobacter immobilis	ATCC 15934	AimI	?	?	?	?	?	?	66
Achromobacter species 697	C. Kessler	Asp697 (AvaII)	GG(A/T)CC	35	73	6	1	8	143
Achromobacter species 700	C. Kessler	Asp700 (XmnI)	GAANN↑NNTTC	24	5	0	3	2	143
Achromobacter species 703	C. Kessler	Asp703 (XhoI)	CTCGAG	1	6	0	1	0	143
Achromobacter species 707	C. Kessler	Asp707 (ClaI)	ATCGAT	15	2	0	0	1	143
Achromobacter species 708	C. Kessler	Asp708 (PstI)	CTGCAG	28	30	2	1	1	143
Achromobacter species 718	C. Kessler	Asp718 (KpnI)	G↑GTACC	2	8	1	0	0	19
Achromobacter species 742	C. Kessler	Asp742I (HaeIII)	GGCC	149	216	18	11	22	381

Microorganism	Source	Enzyme	Sequence	Number of cleavage sites					References
				λ	Ad2	SV40	ΦX	pBR	
Achromobacter species 748	C. Kessler	AspP748I (HpaII)	CCGG	328	171	1	5	26	381
Achromobacter species 763	C. Kessler	Asp763I (ScaI)	AGTACT	5	5	0	0	1	381
Acinetobacter calcoaceticus	R.J. Roberts	AccI	GT↑(A/C)(G/T)AC	9	17	1	2	2	342
		AccII (FnuDII)	CG↑CG	157	303	0	14	23	342;150
		AccIII (BspMII)	T↑CCGGA	24	8	0	0	1	211,357;357
Acinetobacter calcoaceticus EBF 65/65	A. Vivian	AccEBI (BamHI)	G↑GATCC	5	3	1	0	1	367
Actinomadura madurae	ATCC 15904	AmaI (NruI)	TCGCGA	5	5	0	2	1	116
Actinosynnema pretiosum	ATCC 31281	AprI (NaeI)	GCCGGC	1	13	1	0	4	361
Agmenellum quadruplicatum	W.F. Doolittle	AquI (AvaI)	CPyCGPuG	8	40	0	1	1	163
Agrobacterium tumefaciens	ATCC 15955	AtuAI	?	>30	>30	?	?	?	260
Agrobacterium tumefaciens	IAM B-26-1	AtuIAMI	?	?	?	?	?	?	290
Agrobacterium tumefaciens B6806	E. Nester	AtuBI (EcoRII)	CC(A/T)GG	71	136	17	2	6	241
Agrobacterium tumefaciens C58	E. Nester	AtuCI (BclI)	TGATCA	8	5	1	0	0	260
Agrobacterium tumefaciens ID 135	C. Kado	AtuII (EcoRII)	CC(A/T)GG	71	136	17	2	6	170
Agrobacterium tumefaciens IIBV7	G. Roizes	AtuBVI	?	>14	?	1	0	?	240
Alcaligenes species	N. Brown	AspAI (BstEII)	G↑GTNACC	13	10	0	0	0	27
Alcaligenes species 47	C. Kessler	Asp47I (XhoI)	CTCGAG	1	6	0	1	0	381
Alcaligenes species 52	C. Kessler	Asp52I (HindIII)	AAGCTT	6	12	6	0	0	381
Alcaligenes species 78	C. Kessler	Asp78I (StuI)	AGGCCT	6	11	7	1	0	381

Microorganism	Source	Enzyme	Sequence	Number of cleavage sites					References
				λ	Ad2	SV40	ΦX	pBR	
Alcaligenes species RFL 36	A.A. Janulaitis	Asp36I (PstI)	CTGCAG	28	30	2	1	1	122
Alteromonas putrefaciens	M. Sargent	ApuI (Sau96I)	GGNCC	74	164	11	2	15	368
Anabaena catanula	CCAP 1403/1	AcaI	?	?	?	?	?	?	114
Anabaena cylindrica	CCAP 1403/2a	AcyI	GPu↑CGPyC	40	44	0	7	6	55
Anabaena flos-aquae	CCAP 1403/13f	AflI (AvaII)	G↑G(A/T)CC	35	73	6	1	8	328
		AflII	C↑TTAAG	3	4	1	2	0	328
		AflIII	A↑CPuPyGT	20	25	0	2	1	328
Anabaena oscillarioides	CCAP 1403/11	AosI (MstI)	TGC↑GCA	15	17	0	1	4	56
		AosII (AcyI)	GPu↑CGPyC	40	44	0	7	6	56
Anabaena species	CCAP 1403/9	AocI (SauI)	CC↑TNAGG	2	7	0	0	0	53
		AocII	?	?	?	?	?	?	53
Anabaena strain Waterbury	ATCC 29208	AstWI (AcyI)	GPu↑CGPyC	40	44	0	7	6	54
Anabaena subcylindrica	CCAP 1403/4b	AsuI	G↑GNCC	74	164	11	2	15	113
		AsuII	TT↑CGAA	7	1	0	0	0	211,54,54
		AsuIII (AcyI)	GPu↑CGPyC	40	44	0	7	6	54
Anabaena variabilis	ATCC 27892	AvaI	C↑PyCGPuG	8	40	0	1	1	208;115
		AvaII	G↑G(A/T)CC	35	73	6	1	8	208;296,115,71
		AvaIII	ATGCAT	14	9	3	0	0	239;239,268
Anabaena variabilis uw	E.C. Rosenvold	AvrI (AvaI)	CPyCGPuG	8	40	0	1	1	243
		AvrII	C↑CTAGG	2	2	2	0	0	243;242
Aphanothece halophytica	ATCC 29534	AhaI (CauII)	CC(C/G)GG	114	97	0	1	10	327

Microorganism	Source	Enzyme	Sequence	Number of cleavage sites					References
				λ	Ad2	SV40	ΦX	pBR	
Arthrobacter luteus	ATCC 21606	AhaII (AcyI)	GPu↑CGPyC	40	44	0	7	6	327;27,254
Arthrobacter nicotianae	ATCC 15236	AhaIII	TTT↑AAA	13	12	12	2	3	326
Arthrobacter pyridinolis	R. DiLauro	AluI	AG↑CT	143	158	34	24	16	236;236;336,376;337,376
Azospirillum amazonense	G. Schwabe	AnII (NaeI)	GCCGGC	1	13	1	0	4	202
Azospirillum brasilense	ATCC 29711	ApyI (BstNI)	CC↑(A/T)GG	71	136	17	2	6	58
Bacillus acidocaldarius	ATCC 27009	AamI	?	?	?	?	?	?	259
Bacillus alcalophilus 36	M.V. Jones	AbrI (XhoI)	C↑TCGAG	1	6	0	1	0	259
Bacillus alvei	ATCC 6348	BacI (SacII)	CCGCGG	4	33	0	1	0	190,211
Bacillus amyloliquefaciens F	ATCC 23350	Bac36I (Sau96I)	G↑GNCC	74	164	11	2	15	368
Bacillus amyloliquefaciens H	F.E. Young	BavI (PvuII)	CAG↑CTG	15	24	3	0	1	192
Bacillus amyloliquefaciens K	T. Kaneko	BamFI (BamHI)	GGATCC	5	3	1	0	1	267
Bacillus amyloliquefaciens N	T. Ando	BamHI	G↑GATCC	5	3	1	0	1	329;238;103;103
		BamKI (BamHI)	GGATCC	5	3	1	0	1	267
		BamNI (BamHI)	GGATCC	5	3	1	0	1	266
Bacillus aneurinolyticus	IAM 1077	BamNx (AvaII)	G↑G(A/T)CC	35	73	6	1	8	265,266;119
		BanI (HgiCI)	G↑GPyPuCC	25	57	1	3	9	289;289,250
		BanII (HgiJII)	GPuGCPy↑C	7	57	2	0	2	289
		BanIII (ClaI)	AT↑CGAT	15	2	0	0	1	289
Bacillus brevis	ATCC 9999	BbvI	GCAGC (8/12)	199	179	22	14	21	85;83,250;103;103
Bacillus brevis 80	V.M. Kramarov	BbvII	GAAGAC (2/6)	24	27	3	3	3	187
Bacillus brevis S	A.P. Zarubina	BbvSI	GC(A/T)GC	specific methylase					314

Microorganism	Source	Enzyme	Sequence	Number of cleavage sites λ	Ad2	SV40	ΦX	pBR	References
Bacillus caldolyticus	A. Atkinson	BclI	T↑GATCA	8	5	1	0	0	15
Bacillus centrosporus	A.A. Janulaitis	BcnI (CauII)	CC↑(C/G)GG	114	97	0	1	10	129,130;124
Bacillus cereus	IAM 1229	Bce1229	?	>10	?	?	?	?	267
Bacillus cereus	ATCC 14579	Bce14579	?	>10	?	?	?	?	267
Bacillus cereus	T. Ando	Bce170 (PstI)	CTGCAG	28	30	2	1	1	267
Bacillus cereus	IOC 243	Bce243 (Sau3AI)	↑GATC	116	87	8	0	22	52
Bacillus cereus	ATCC 31293	BceFI (FnuDII)	CGCG	157	303	0	14	23	219
Bacillus cereus Rf sm st	T. Ando	BceR (FnuDII)	CGCG	157	303	0	14	23	267
Bacillus globigii	G.A. Wilson	BglI	GCCNNNN↑NGGC	29	20	1	0	3	60,330;11,310
		BglII	A↑GATCT	6	11	0	0	0	60,330;224
Bacillus megaterium	J. Upcroft	BmeI	?	>10	>20	4	?	?	78
Bacillus megaterium 216	V.M. Kramarov	Bme216 (AvaII)	GG(A/T)CC	35	73	6	1	8	155
Bacillus megaterium 899	B899	Bme899	?	>5	?	?	?	?	267
Bacillus megaterium B205-3	T. Kaneko	Bme205	?	>10	?	?	?	?	267
Bacillus pumilus AHU1387A	T. Ando	BpuI	?	6	>30	2	?	?	118
Bacillus species	P. Eastlake	BscI (ClaI)	AT↑CGAT	15	2	0	0	1	380;27
Bacillus species H	NEB 394	BspHI	TCATGA	8	3	2	3	4	388
Bacillus species M	NEB 356	BspMI	ACCTGC (4/8)	41	39	0	3	1	202,203
		BspMII	T↑CCGGA	24	8	0	0	1	202,203
Bacillus sphaericus	IAM 1286	Bsp1286 (SduI)	G(A/G/T)GC(A/C/T)↑C	38	105	4	3	10	267;211,252
Bacillus sphaericus JL14	R. Mullings	BspBI (PstI)	CTGCA↑G	28	30	2	1	1	365

Microorganism	Source	Enzyme	Sequence	Number of cleavage sites				References
				λ	Ad2	SV40	ΦX	pBR	
Bacillus sphaericus JL4B	R. Mullings	BspBII (AsuI)	G↑GNCC	74	164	11	2	15	365
		BspAI (MboI)	↑GATC	116	87	8	0	22	365
Bacillus sphaericus R	P. Venetianer	BspRI (HaeIII)	GG↑CC	149	216	18	11	22	149;316;153
Bacillus stearothermophilus	D. Comb	BstNI (EcoRII)	CC↑(A/T)GG	71	136	17	2	6	253
Bacillus stearothermophilus ATCC 12980		BstPI (BstEII)	G↑GTNACC	13	10	0	0	0	226
Bacillus stearothermophilus 1503-4R	N. Welker	BstI (BamHI)	G↑GATCC	5	3	1	0	1	38;41
Bacillus stearothermophilus 240	A. Atkinson	BstAI	?	?	?	?	?	?	17
Bacillus stearothermophilus C1	N. Welker	BstCI (HaeIII)	GGCC	149	216	18	11	22	162
Bacillus stearothermophilus C11	N. Welker	BssCI (HaeIII)	GGCC	149	216	18	11	22	162
Bacillus stearothermophilus ET	N. Welker	BstEI	?	?	?	?	?	?	194
		BstEII	G↑GTNACC	13	10	0	0	0	194;165
		BstEIII (MboI)	GATC	116	87	8	0	22	194;88;211
Bacillus stearothermophilus G3	N. Welker	BstGI (BclI)	TGATCA	8	5	1	0	0	162
		BstGII (EcoRII)	CC(A/T)GG	71	136	17	2	6	162
Bacillus stearothermophilus G6	N. Welker	BssGI (BstXI)	CCANNNNNTGG	13	10	1	3	0	162
		BssGII (MboI)	GATC	116	87	8	0	22	162
Bacillus stearothermophilus H1	N. Welker	BstHI (XhoI)	CTCGAG	1	6	0	1	0	162
Bacillus stearothermophilus H3	N. Welker	BssHI (XhoI)	CTCGAG	1	6	0	1	0	162

Microorganism	Source	Enzyme	Sequence	Number of cleavage sites				References
				λ	Ad2	SV40	ΦX	pBR	
Bacillus stearothermophilus H4	N. Welker	BssHIII (BsePI)	G↑CGCGC	6	52	0	1	0	162;255
Bacillus stearothermophilus NUB 36	N. Welker	BsrHI (BsePI)	GCGCGC	6	52	0	1	0	162
Bacillus stearothermophilus NUB31	N. Welker	BsmI	GAATGC (1/−1)	46	10	4	3	1	211
Bacillus stearothermophilus P1	N. Welker	Bst31I (BstEII)	GGTNACC	13	10	0	0	0	105
Bacillus stearothermophilus P5	N. Welker	BssPI	?	>30	?	?	?	?	162
Bacillus stearothermophilus P6	N. Welker	BsrPI	?	11	>20	?	0	0	162
Bacillus stearothermophilus P8	N. Welker	BsrPII (Sau3AI)	GATC	116	87	8	0	22	162
Bacillus stearothermophilus P9	N. Welker	BsePI	GCGCGC	6	52	0	1	0	162
Bacillus stearothermophilus T12	N. Welker	BsaPI (MboI)	GATC	116	87	8	0	22	162
Bacillus stearothermophilus	N. Welker	BsoPI (BsePI)	GCGCGC	6	52	0	1	0	162
Bacillus stearothermophilus	N. Welker	BstTI (BstXI)	CCANNNNNTGG	13	10	1	3	0	162
Bacillus stearothermophilus V	C. Vasquez	BstVI (XhoI)	CTCGAG	1	6	0	1	0	382
Bacillus stearothermophilus X1	N. Welker	BstXI	CCANNNNN↑NTGG	13	10	1	3	0	162;256
Bacillus stearothermophilus		BstXII (MboI)	GATC	116	87	8	0	22	162
Bacillus stearothermophilus strain 822	T. Oshima	BseI (HaeIII)	GGCC	149	216	18	11	22	271
		BseII (HpaI)	GTTAAC	14	6	4	3	0	271
Bacillus subtilis	IAM 1076	Bsu1076 (HaeIII)	GGCC	149	216	18	11	22	267

Microorganism	Source	Enzyme	Sequence	λ	Ad2	SV40	ΦX	pBR	References
Bacillus subtilis	IAM 1114	Bsu1114 (HaeIII)	GGCC	149	216	18	11	22	267
Bacillus subtilis	ATCC 14593	Bsu1145	?	>20	?	?	?	?	267
Bacillus subtilis	IAM 1192	Bsu1192I (HpaII)	CCGG	328	171	1	5	26	267;228
		Bsu1192II (FnuDII)	CGCG	157	303	0	14	23	228
Bacillus subtilis	IAM 1193	Bsu1193 (FnuDII)	CGCG	157	303	0	14	23	267;228
Bacillus subtilis	IAM 1259	Bsu1259	?	>8	?	?	?	?	267
Bacillus subtilis	ATCC 6633	Bsu6633 (FnuDII)	CGCG	157	303	0	14	23	267;120
Bacillus subtilis	IAM 1247	BsuBI (PstI)	CTGCAG	28	30	2	1	1	267;110
Bacillus subtilis	IAM 1231	BsuEII (FnuDII)	CGCG	157	303	0	14	23	228;228;136
		BsuFI (HpaII)	*CCGG	328	171	1	5	26	267;228;136
Bacillus subtilis Marburg 168	T. Ando	BsuM (XhoI)	CTCGAG	1	6	0	1	0	267;136;136
Bacillus subtilis strain R	T. Trautner	BsuRI (HaeIII)	GG↑CC	149	216	18	11	22	26;25;95
Bacillus thuringiensis	R.R. Azizbekyan	BtI (AvaII)	GG(A/T)CC	35	73	6	1	8	4
Bacillus vulgatis	OSB816	BvuI (HgiJIII)	GPuGCPy↑C	7	57	2	0	2	8
Bifidobacterium bifidum YIT4007	T. Khosaka	BbI (PstI)	CTGCAG	28	30	2	1	1	147
		BbII (AcyI)	GPu↑CGPyC	40	44	0	7	6	147
		BbIII (XhoI)	CTCGAG	1	6	0	1	0	147
		BbIV	?	?	?	0	0	0	145
Bifidobacterium breve S1	ATCC 15700	BbeSI	?	?	?	?	?	?	145
Bifidobacterium breve S50	ATCC 15698	BbeAI (NarI)	GGCGCC	1	20	0	2	4	145
		BbeAII	?	?	?	?	?	?	145

Microorganism	Source	Enzyme	Sequence	Number of cleavage sites					References
				λ	Ad2	SV40	ΦX	pBR	
Bifidobacterium breve YIT4006	H. Takahashi	BbeI (NarI)	GGCGC↑C	1	20	0	2	4	148
		BbeII	?	?	?	?	?	?	145
Bifidobacterium infantis 659	ATCC 25962	BinI	GGATC (4/5)	58	35	6	0	12	146
Bifidobacterium infantis S76e	ATCC 15702	BinSI (EcoRII)	CC(A/T)GG	71	136	17	2	6	145
		BinSII (NarI)	GGCGCC	1	20	0	2	4	145
Bifidobacterium longum E194b	ATCC 15707	BloI	?	?	?	?	?	?	145
Bifidobacterium thermophilum RU326	ATCC 25866	BthI (XhoI)	CTCGAG	1	6	0	1	0	145
		BthII (BinI)	GGATC	58	35	6	0	12	145
Bordetella bronchiseptica	ATCC 19395	BbrI (HindIII)	AAGCTT	6	12	6	0	1	211
Bordetella pertussis	P. Novotny	BpeI (HindIII)	AAGCTT	6	12	6	0	1	90,141
Brevibacterium albidum	ATCC 15831	BalI	TGG↑CCA*	18	17	0	0	1	79;79;306
Brevibacterium luteum	ATCC 15830	BluI (XhoI)	C↑TCGAG	1	6	0	1	0	84
		BluII (HaeIII)	GGCC	149	216	18	11	22	311
Brevibacterium protophormiae	IFO 12128	BprI	?	?	?	?	?	?	290
Calothrix scopulorum	CCAP 1410/5	CscI (SacII)	CCGC↑GG	4	33	0	1	0	62
Caryophanon latum	ATCC 15219	ClmI (HaeIII)	GGCC	149	216	18	11	22	279
		ClmII (AvaII)	GG(A/T)CC	35	73	6	1	8	279
Caryophanon latum	DSM 484	ClI (HaeIII)	GG↑CC	149	216	18	11	22	190
Caryophanon latum H7	W.C. Trentini	CaI	?	14	?	?	?	?	190
Caryophanon latum L	H. Mayer	ClaI	AT↑CGAT	15	2	0	0	1	188
Caryophanon latum RII	H. Mayer	ClaI	?	>20	?	?	?	?	190

Microorganism	Source	Enzyme	Sequence	Number of cleavage sites					References
				λ	Ad2	SV40	ΦX	pBR	
Caulobacter crescentus CB-13	R.J. Syddall	CcrII (XhoI)	CTCGAG	1	6	0	1	0	297
Caulobacter crescentus CB-13Bla	J. Poindexter	CcrI (XhoI)	C↑TCGAG	1	6	0	1	0	297
Caulobacter fusiformis	A.A. Janulaitis	CfuI (DpnI)	$\overset{*}{G}ATC$	cleaves methylated DNA					125
Cellulomonas flavigena	IFO 3753	CflI (PstI)	CTGCA↑G	28	30	2	1	1	356
Chloroflexus aurantiacus	A. Bingham	CauI (AvaII)	G↑G(A/T)CC	35	73	6	1	8	16;201
		CauII	CC↑(C/G)GG	114	97	0	1	10	16;173,201
		CauIII (PstI)	CTGCAG	28	30	2	1	1	10
Chromatium vinosum	G.C. Grosveld	CvnI (SauI)	CC↑TNAGG	2	7	0	0	0	94
Chromobacterium violaceum ATCC 12472		CviI	?	?	?	?	?	?	66
Citrobacter diversus RFL27	A.A. Janulaitis	Cdi27I (EcoRII)	CC(A/T)GG	71	136	17	2	6	122
Citrobacter freundii RFL10	A.A. Janulaitis	Cfr10I	Pu↑CCGGPy	61	40	1	0	7	131,132,134;379
Citrobacter freundii RFL11	A.A. Janulaitis	Cfr11I (EcoRII)	CC(A/T)GG	71	136	17	2	6	132,134
Citrobacter freundii RFL13	A.A. Janulaitis	Cfr13I (AsuI)	G↑G$\overset{*}{N}$CC	74	164	11	2	15	18,132
Citrobacter freundii RFL14	A.A. Janulaitis	Cfr14I (CfrI)	Py↑GGCCPu	39	70	0	2	6	132
Citrobacter freundii RFL2	A.A. Janulaitis	CfrI	Py↑GGCCPu	39	70	0	2	6	133
Citrobacter freundii RFL4	A.A. Janulaitis	Cfr4I (AsuI)	GGNCC	74	164	11	2	15	132,134
Citrobacter freundii RFL5	A.A. Janulaitis	Cfr5I (EcoRII)	CC(A/T)GG	71	136	17	2	6	132,134
Citrobacter freundii RFL6	A.A. Janulaitis	Cfr6I (PvuII)	CAGCTG	15	24	3	0	1	132,134
Citrobacter freundii RFL7	A.A. Janulaitis	Cfr7I (BstEII)	GGTNACC	13	10	0	0	0	132
Citrobacter freundii RFL8	A.A. Janulaitis	Cfr8I (AsuI)	GGNCC	74	164	11	2	15	132,134

Microorganism	Source	Enzyme	Sequence	Number of cleavage sites				References
				λ	Ad2	SV40	ΦX	pBR	
Citrobacter freundii RFL9	A.A. Janulaitis	Cfr9I (SmaI)	C↑CCGGG	3	12	0	0	0	132,134
Citrobacter freundii S39	W. Piepersberg	Cfr-S37I (EcoRII)	CC(A/T)GG	71	136	17	2	6	381
Clostridium formicoaceticum	ATCC 23439	CfoI (HhaI)	GCGC	215	375	2	18	31	182
Clostridium histolyticum	R. Hansen	ChI	?	?	?	?	?	?	100
Clostridium pasteurianum	NRCC 33011	CpaI (MboI)	GATC	116	87	8	0	22	316
Clostridium perfringens	R. Hansen	CpfI (Sau3AI)	↑GATC	116	87	8	0	22	100
Corynebacterium humiferum	ATCC 21108	ChuI (HindIII)	AAGCTT	6	12	6	0	1	66
		ChuII (HindII)	GTPyPuAC	35	25	7	13	2	66
Corynebacterium petrophilum	ATCC 19080	CpeI (BclI)	TGATCA	8	5	1	0	0	68
Cystobacter velatus Plv9	H. Reichenbach	CveI	?	?	?	?	?	?	190
Dactylococcopsis salina	A.E. Walsby	DsaI	C↑CPuPyGG	46	82	3	3	2	364
		DsaII (HaeIII)	GG↑CC	149	216	18	11	22	364
Deinococcus radiophilus	ATCC 27603	DraI (AhaIII)	TTT↑AAA	13	12	12	2	3	227
		DraII	PuG↑GNCCPy	3	44	3	0	4	100,93;57,93
		DraIII	CACNNN↑GTG	10	10	0	1	0	100,93;57,93
Desulfococcus mobilus	W. Zillig	DmoI	?	?	?	?	?	?	193
Desulfovibrio desulfuricans	ATCC 27774	DdsI (BamHI)	GGATCC	5	3	1	0	1	181
Desulfovibrio desulfuricans Norway strain	H. Peck	DdeI	C↑TNAG	104	97	20	14	8	182;80
		DdeII (XhoI)	CTCGAG	1	6	0	1	0	211
Diplococcus pneumoniae	S. Lacks	DpnI	GA↑TC*	cleaves methylated DNA					159;76,160
		DpnII (MboI)	GATC	116	87	8	0	22	159;160

Microorganism	Source	Enzyme	Sequence	λ	Ad2	SV40	ΦX	pBR	References
Enterobacter aerogenes	P.R. Whitehead	EaeI (CfrI)	Py↑GGCCPu*	39	70	0	2	6	327
Enterobacter aerogenes	ATCC 15038	EaePI (PstI)	CTGCAG	28	30	2	1	1	219
Enterobacter agglomerans	NEB 368	EagI (XmaIII)	C↑GGCCG	2	19	0	0	1	355
Enterobacter cloacae	DSM 30056	EcaI (BstEII)	G↑GTNACC	13	10	0	0	0	109
Enterobacter cloacae		EcaII (EcoRII)	CC(A/T)GG	71	136	17	2	6	211
Enterobacter cloacae	DSM 30060	EccI (SacII)	CCGCGG	4	33	0	1	0	189;211
Enterobacter cloacae	H. Hartmann	EclI	?	14	?	?	?	?	101
Enterobacter cloacae		EclII (EcoRII)	CC(A/T)GG	71	136	17	2	6	101
Enterobacter cloacae	W. Piepersberg	EclS39I (EcoRII)	CC(A/T)GG	71	136	17	2	6	381
Enterobacter cloacae 593	C. Kessler	Ecl593I (PstI)	CTGCAG	28	30	2	1	1	381
Enterobacter cloacae RFL28	A.A. Janulaitis	Ecl28I (SacII)	CCGCGG	4	33	0	1	0	122
Erwinia rhaponici	D. Jones	ErpI (AvaII)	G↑G(A/T)CC	35	73	6	1	8	368
Escherichia coli	S. Glover	EcoR124	GAANNNNNNPuTCG	Type I enzyme					387
Escherichia coli		EcoR124/3	GAANNNNNNNPuTCG	Type I enzyme					387
Escherichia coli (PI)	K. Murray	EcoPI	AGACC*	Type III enzyme					96;5;23;24;5,102
Escherichia coli 15T-	T. Bickle	EcoA	GAGNNNNNNNGTCA*	Type I enzyme					293,156
Escherichia coli 2bT	ICR 0020	EcoICRI (SacI)	GAGCTC*	2	16	0	0	0	290
Escherichia coli B	W. Arber	EcoB	TGANNNNNNNNTGCT	Type I enzyme					67;166,231;167;313
Escherichia coli CK	S.S. Debov	EcoCK	?	4	?	?	?	?	308
Escherichia coli E1585-68	M. Yoshikawa	EcoVIII (HindIII)	A↑AGCTT	6	12	6	0	1	198
Escherichia coli E166	N.E. Murray	EcoD	TTANNNNNNNGTCPy	Type I enzyme					212

Microorganism	Source	Enzyme	Sequence	Number of cleavage sites					References
				λ	Ad2	SV40	ΦX	pBR	
Escherichia coli H709c	I. Orskov	EcoO109 (DraII)	PuG↑GNCCPy	3	44	3	0	4	200
Escherichia coli J62 pLG74	L.I. Glatman	EcoRV	GAT↑ATC	21	9	1	0	1	144;144,251
Escherichia coli K	M. Meselson	EcoK	AACNNNNNNGTGC	Type I enzyme					195;12,139;97
Escherichia coli P15	W. Arber	EcoP15	CAGCAG	Type III enzyme					233;98
Escherichia coli R245	R.N. Yoshimori	EcoRII	↑CC(A/T)GG	71	136	17	2	6	340;13,22;340;22
Escherichia coli RFL24	A.A. Janulaitis	Eco24I (HgiJII)	GPuGCPyC	7	57	2	0	2	122
Escherichia coli RFL25	A.A. Janulaitis	Eco25I (HgiJII)	GPuGCPyC	7	57	2	0	2	122
Escherichia coli RFL26	A.A. Janulaitis	Eco26I (HgiJII)	GPuGCPyC	7	57	2	0	2	134
Escherichia coli RFL31	A.A. Janulaitis	Eco31I	GGTCTC (1/5)	2	18	0	0	1	384
Escherichia coli RFL32	A.A. Janulaitis	Eco32I (EcoRV)	GATATC	21	9	1	0	1	134
Escherichia coli RFL35	A.A. Janulaitis	Eco35I (HgiJII)	GPuGCPyC	7	57	2	0	2	122
Escherichia coli RFL38	A.A. Janulaitis	Eco38I (EcoRII)	CC(A/T)GG	71	136	17	2	6	121
Escherichia coli RFL39	A.A. Janulaitis	Eco39I (AsuI)	GGNCC	74	164	11	2	15	121
Escherichia coli RFL40	A.A. Janulaitis	Eco40I (HgiJII)	GPuGCPyC	7	57	2	0	2	121
Escherichia coli RFL41	A.A. Janulaitis	Eco41I (HgiJII)	GPuGCPyC	7	57	2	0	2	121
Escherichia coli RFL47	A.A. Janulaitis	Eco47I (AvaII)	GG(A/T)CC	35	73	6	1	8	128
		Eco47II (AsuI)	GGNCC	74	164	11	2	15	128
		Eco47III	AGC↑GCT	2	13	1	0	4	128
Escherichia coli RFL48	A.A. Janulaitis	Eco48I (PstI)	CTGCAG	28	30	2	1	1	127
Escherichia coli RFL49	A.A. Janulaitis	Eco49I (PstI)	CTGCAG	28	30	2	1	1	127
Escherichia coli RFL50	A.A. Janulaitis	Eco50I (HgiCI)	GGPyPuCC	25	57	1	3	9	127

Microorganism	Source	Enzyme	Sequence	λ	Ad2	SV40	ΦX	pBR	References
Escherichia coli RFL51	A.A. Janulaitis	Eco51I (Eco31I)	?	?	?	?	?	?	127
		Eco51II (CauII)	CC(C/G)GG	114	97	0	1	10	127
Escherichia coli RFL52	A.A. Janulaitis	Eco52I (XmaIII)	CGGCCG	2	19	0	0	1	134
Escherichia coli RFL55	A.A. Janulaitis	Eco55I (SacII)	CCGCGG	4	33	0	1	0	127
Escherichia coli RFL56	A.A. Janulaitis	Eco56I (NaeI)	GCCGGC	1	13	1	0	4	134
Escherichia coli RFL60	A.A. Janulaitis	Eco60I (EcoRII)	CC(A/T)GG	71	136	17	2	6	134
Escherichia coli RFL61	A.A. Janulaitis	Eco61I (EcoRII)	CC(A/T)GG	71	136	17	2	6	134
Escherichia coli RY13	R.N. Yoshimori	EcoRI	G↑AA*TTC	5	5	1	0	1	91;104;91;59
		EcoRI'	PuPuA↑TPyPy	190	100	24	16	15	207
Escherichia coli TB104	N. Terakado	EcoT104 (StyI)	CC(A/T)(A/T)GG	10	44	8	0	1	338
Escherichia coli TB14	N. Terakado	EcoT14I (StyI)	C↑C(A/T)(A/T)GG	10	44	8	0	1	338
Escherichia coli TB22	N. Terakado	EcoT22 (AvaIII)	ATGCA↑T	14	9	3	0	0	338
Escherichia coli pDXX1	A. Piekarowicz	EcoDXXI	TCANNNNNNNATTC	Type I enzyme					385;391
Eucapsis species	PCC 6906	EspI	GC↑TNAGC	7	8	1	0	0	34
Fischerella species	ATCC 29114	FspI (MstI)	TGCGCA	15	17	0	1	4	298
		FspII (AsuII)	TT↑CGAA	7	1	0	0	0	298
Fischerella species	M. Sargent	FspMSI (AvaII)	G↑G(A/T)CC	35	73	6	1	8	368
Flavobacterium balustinum	ATCC 33487	FbaI (BclI)	TGATCA	8	5	1	0	0	359
Flavobacterium breve	NEB 379	FbrI (Fnu4HI)	GC↑NGC	379	411	24	31	42	359
Flavobacterium indologenes	NEB 382	FinI	GTCCC	38	59	8	2	4	202
		FinII (HpaII)	CCGG	328	171	1	5	26	202

Microorganism	Source	Enzyme	Sequence	λ	Ad2	SV40	ΦX	pBR	References
Flavobacterium indoltheticum	ATCC 27950	FinSI (HaeIII)	GGCC	149	216	18	11	22	359
Flavobacterium okeanokoites	IFO 12536	FokI	GGATG (9/13)	150	78	11	8	12	288
Flavobacterium species	NEB 380	FspMI (FnuDII)	CGCG	157	303	0	14	23	202
Flavobacterium suaveolens	ATCC 13718	FsuI (Tth111I)	GACNNNGTC	2	12	0	0	1	390
Fremyella diplosiphon	PCC 7601	FdI (AvaII)	G↑G(A/T)CC	35	73	6	1	8	309,286
		FdII (MstI)	TGC↑GCA	15	17	0	1	4	309,286
Fusobacterium nucleatum 48	M. Smith	Fnu48I	?	>50	?	?	>10	?	176
Fusobacterium nucleatum 4H	M. Smith	Fnu4HI	GC↑NGC	379	411	24	31	42	172
Fusobacterium nucleatum A	M. Smith	FnuAI (HinfI)	G↑ANTC	148	72	10	21	10	177
		FnuAII (MboI)	↑GATC	116	87	8	0	22	177,211
Fusobacterium nucleatum C	M. Smith	FnuCI (MboI)	↑GATC	116	87	8	0	22	177
Fusobacterium nucleatum D	M. Smith	FnuDI (HaeIII)	GG↑CC	149	216	18	11	22	177
		FnuDII	CG↑CG	157	303	0	14	23	177
		FnuDIII (HhaI)	GCG↑C	215	375	2	18	31	177
Fusobacterium nucleatum E	M. Smith	FnuEI (Sau3AI)	↑GATC	116	87	8	0	22	177
Gluconobacter albidus	IFO 3251	GaII (SacII)	CCGC↑GG	4	33	0	1	0	356
Gluconobacter cerinus	IFO 3285	GceGLI (SacII)	CCGC↑GG	4	33	0	1	0	348
Gluconobacter cerinus	IFO 3262	GceI (SacII)	CCGC↑GG	4	33	0	1	0	356
Gluconobacter dioxyacetonicus	IAM 1814	GdiI (StuI)	AGG↑CCT	6	11	7	1	0	311
		GdiII	PyGGCCG (−5/−1)	21	53	0	2	5	311
Gluconobacter dioxyacetonicus	IAM 1840	GdoI (BamHI)	GGATCC	5	3	1	0	1	262

Microorganism	Source	Enzyme	Sequence	λ	Ad2	SV40	ΦX	pBR	References
Gluconobacter gluconicus	IFO 3285	GgII	?	?	?	?	?	?	290
Gluconobacter industricus	IFO 3260	GinI (BamHI)	GGATCC	5	3	1	0	1	290
Gluconobacter oxydans sub. melonogenes	IAM 1836	GoxI (BamHI)	GGATCC	5	3	1	0	1	262
Gluconobacter suboxydans H-15M	M.S. Loytsianskaya	GsbI (GsuI)	CTCCAG	25	32	6	3	4	123
Gluconobacter suboxydans H-15T	M.S. Loytsianskaya	GsuI	CTCCAG	25	32	6	3	4	123
Haemophilus aegyptius	ATCC 11116	HaeI	(A/T)GG↑CC(A/T)	64	56	11	6	7	213
		HaeII	PuGCGC↑Py	48	76	1	8	11	235;307
		HaeIII	GG↑CC*	149	216	18	11	22	196;25;184;184
Haemophilus aphrophilus	ATCC 19415	HapI	?	>30	?	?	?	?	211
		HapII (HpaII)	C↑CGG	328	171	1	5	26	301;291
Haemophilus gallinarum	ATCC 14385	HgaI	GACGC (5/10)	102	87	0	14	11	301;30;287
Haemophilus haemoglobinophilus	ATCC 19416	HhgI (HaeIII)	GGCC	149	216	18	11	22	211
Haemophilus haemolyticus	ATCC 10014	HhaI	GCG↑C*	215	375	2	18	31	237;237;185
		HhaII (HinfI)	G↑ANTC*	148	72	10	21	10	183;183;276
Haemophilus influenzae 1056	J. Stuy	Hin1056I (FnuDII)	CGCG	157	303	0	14	23	217
		Hin1056II	?	>30	>30	0	5	?	217
Haemophilus influenzae 173	J. Chirikjian	Hin173 (HindIII)	AAGCTT	6	12	6	0	1	280
Haemophilus influenzae GU	J. Chirikjian	HinGUI (HhaI)	GCGC	215	375	2	18	31	280;40
		HinGUII (FokI)	GGATG	150	78	11	8	12	280;213,303

Microorganism	Source	Enzyme	Sequence	λ	Ad2	SV40	ΦX	pBR	References
Haemophilus influenzae H-1	M. Takanami	HinHI (HaeII)	PuGCGCPy	48	76	1	8	11	301
Haemophilus influenzae JC9	A. Piekarowicz	HinJCI (HindII)	GTPy↑PuAC	35	25	7	13	2	223
		HinJCII (HindIII)	AAGCTT	6	12	6	0	1	223
Haemophilus influenzae P1	S. Shen	HinPII (HhaI)	G↑CGC	215	375	2	18	31	264
Haemophilus influenzae Rb	C.A. Hutchison	HinbIII (HindIII)	AAGCTT	6	12	6	0	1	197,211
Haemophilus influenzae Rc	A. Landy, G. Leidy	HincII (HindII)	GTPyPuAC	35	25	7	13	2	161
Haemophilus influenzae Rd (exo-mutant)	S. H. Goodgal	HindI	$\overset{*}{C}AC$	specific methylase					244;245
		HindII	GTPy↑PuAC*	35	25	7	13	2	278;142;244;245
		HindIII	A↑$\overset{*}{A}$GCTT	6	12	6	0	1	216;216;244;245
		HindIV	$\overset{*}{G}AC$	specific methylase					244;245
Haemophilus influenzae Rf	C.A. Hutchison	HinfI	G↑ANTC	148	72	10	21	10	197;117,209
		HinfII (HindIII)	AAGCTT	6	12	6	0	1	185
		HinfIII	CGAAT	Type III enzyme					140;222
Haemophilus influenzae S1	S. Shen	HinS1 (HhaI)	GCGC	215	375	2	18	31	264
Haemophilus influenzae S2	S. Shen	HinS2 (HhaI)	GCGC	215	375	2	18	31	264
Haemophilus influenzae serotype b, 1076	J. Stuy	Hin1076III (HindIII)	AAGCTT	6	12	6	0	1	217
Haemophilus influenzae serotype c, 1160	J. Stuy	Hin1160II (HindII)	GTPyPuAC	35	25	7	13	2	217
Haemophilus influenzae serotype c, 1161	J. Stuy	Hin1161II (HindII)	GTPyPuAC	35	25	7	13	2	217
Haemophilus influenzae serotype e	A. Piekarowicz	HineI (HinfIII)	CGAAT	Type III enzyme					220

Microorganism	Source	Enzyme	Sequence	λ	Ad2	SV40	ΦX	pBR	References
Haemophilus parahaemolyticus	C.A. Hutchison	HphI	GGTGA (8/7)	168	99	4	9	12	197;151
Haemophilus parainfluenzae	J. Setlow	HpaI	GTT↑AAC	14	6	4	3	0	263;74;1
		HpaII	C↑CGG	328	171	1	5	26	263;74,184;184
Haemophilus suis	ATCC 19417	HsuI (HindIII)	A↑AGCTT	6	12	6	0	1	211
Halococcus acetoinfaciens	IAM 12094	HacI (MboI)	↑GATC	116	87	8	0	22	356
Halococcus agglomeratus	ATCC 25862	HagI	?	?	?	?	?	?	228
Herpetosiphon giganteus HFS101	H. Foster	HgiI	?	?	57	2	0	2	327
		HgiII	GPuGCPy↑C	7	38	0	3	8	327
Herpetosiphon giganteus HP1023	J.H. Parish	HgiAI	G(A/T)GC(A/T)↑C	28	57	1	3	9	29
Herpetosiphon giganteus HP1049	J.H. Parish	HgiHI (HgiCI)	G↑GPyPuCC	25	44	0	7	6	327
		HgiHII (AcyI)	GPu↑CGPyC	40	73	6	1	8	327
		HgiHIII (AvaII)	G↑G(A/T)CC	35	44	0	7	6	327
Herpetosiphon giganteus Hpa1	H. Reichenbach	HgiGI (AcyI)	GPu↑CGPyC	40	44	0	7	6	157
Herpetosiphon giganteus Hpa2	H. Reichenbach	HgiDI (AcyI)	GPu↑CGPyC	40	3	0	0	1	157
Herpetosiphon giganteus Hpg14	H. Reichenbach	HgiDII (SalI)	G↑TCGAC	2	15	?	?	?	157
Herpetosiphon giganteus Hpg24	H. Reichenbach	HgiFI	?	?	73	6	1	8	190
		HgiEI (AvaII)	G↑G(A/T)CC	35	10	1	1	2	157
Herpetosiphon giganteus Hpg32	H. Reichenbach	HgiEII	ACCNNNNNNGGT	14	>20	?	?	?	157
		HgiKI	?	>18	73	6	1	8	190
Herpetosiphon giganteus Hpg5	H. Reichenbach	HgiBI (AvaII)	G↑G(A/T)CC	35					157

Microorganism	Source	Enzyme	Sequence	Number of cleavage sites				References
				λ	Ad2	SV40	ΦX	pBR	
---	---	---	---	---	---	---	---	---	---
Herpetosiphon giganteus Hpg9	H. Reichenbach	HgiCI	↑GGPyPuCC	25	57	1	3	9	157
		HgiCII (AvaII)	G↑G(A/T)CC	35	73	6	1	8	157
		HgiCIII (SalI)	G↑TCGAC	2	3	0	0	1	157
Herpetosiphon giganteus S21	H. Foster	HgiS21I (CauII)	CC(C/G)GG	114	97	0	1	10	381
Klebsiella aerogenes RFL37	A.A. Janulaitis	Kae37I (SacII)	CCGCGG	4	33	0	1	0	121
Klebsiella pneumoniae OK8	J. Davies	KpnI	GGTAC↑C	2	8	1	0	0	275;304
Klebsiella pneumoniae mmK14	W. Piepersberg	KpnK14I (KpnI)	GGTACC	2	8	1	0	0	381
Mastigocladus laminosus CCAP 1447/1		MlaI (AsuII)	TT↑CGAA	7	1	0	0	0	61
Methanococcus aeolicus PL-15/H	K.O. Stetter	MaeI	C↑TAG	13	54	12	3	5	258
		MaeII	A↑CGT	143	83	0	19	10	258
		MaeIII	↑GTNAC	156	118	14	17	17	258
Methanococcus jannashii	H. Escalante	MjaI (MaeI)	CTAG	13	54	12	3	5	344
		MjaII (AsuI)	GGNCC	74	164	11	2	15	344
Microbacterium flavum	IAM 1642	MfII (XhoII)	Pu↑GATCPy	21	22	3	0	8	108
Microbacterium thermosphactum	ATCC 11509	MthI (Sau3AI)	GATC	116	87	8	0	22	162
Micrococcus aurantiacus	IFO 12422	MauI (PstI)	CTGCAG	28	30	2	1	1	290
Micrococcus euryhalis	ATCC 14389	MeuI (MboI)	GATC	116	87	8	0	22	361
Micrococcus kristinae	ATCC 27571	MkrI (PstI)	CTGCAG	28	30	2	1	1	361
Micrococcus luteus	ATCC 540	MleI (BamHI)	GGATCC	5	3	1	0	1	361
Micrococcus luteus	ATCC 400	MltI (AluI)	AG↑CT	143	158	34	24	16	361;362
Micrococcus luteus	IFO 12992	MluI	A↑CGCGT	7	5	0	2	0	288

Microorganism	Source	Enzyme	Sequence	Number of cleavage sites					References
				λ	Ad2	SV40	ΦX	pBR	
Micrococcus radiodurans	ATCC 13939	MraI (SacII)	CCGCGG	4	33	0	1	0	322
Micrococcus species	R. Meagher	MisI (NaeI)	GCCGGC	1	13	1	0	4	194
Micrococcus varians	RFL19	MvaI (BstNI)	CC↑(A/T)GG	71	136	17	2	6	33
Micrococoleus species	D. Comb	MstI	TGC↑GCA	15	17	0	1	4	49;83
		MstII (SauI)	CC↑TNAGG	2	7	0	0	0	250
Micromonospera carbonacea	C. Kessler	McaI (XhoI)	CTCGAG	1	6	0	1	0	143
Moraxella bovis	ATCC 10900	MboI	↑GATC	116	87	8	0	22	77
		MboII	$\overset{*}{\text{GAAGA}}$ (8/7)	130	113	16	11	11	77;28,64;350;350
Moraxella bovis	ATCC 17947	MbvI	?	?	?	?	?	?	138
Moraxella glueidi LG1	J. Davies	MgII	?	?	?	?	?	?	275
Moraxella glueidi LG2	J. Davies	MgIII	?	?	?	?	?	?	275
Moraxella kingae	ATCC 23331	MkiI (HindIII)	AAGCTT	6	12	6	0	1	138
Moraxella nonliquefaciens	ATCC 19996	MnII (HaeIII)	GGCC	149	216	18	11	22	138
		MnIII (HpaII)	CCGG	328	171	1	5	26	138
Moraxella nonliquefaciens	ATCC 17953	MnII	CCTC (7/7)	262	397	51	34	26	341;250
Moraxella nonliquefaciens	ATCC 17954	MnnI (HindII)	GTPyPuAC	35	25	7	13	2	99
		MnnII (HaeIII)	GGCC	149	216	18	11	22	99
		MnnIII	?	>10	>6	3	?	?	99
		MnnIV (HhaI)	GCGC	215	375	2	18	31	99
Moraxella nonliquefaciens	ATCC 19975	MnoI (HpaII)	C↑CGG	328	171	1	5	26	211;7
		MnoII (MnnIII)	?	>10	>6	3	?	?	211

Microorganism	Source	Enzyme	Sequence	Number of cleavage sites				References
				λ	Ad2	SV40	ΦX	pBR	
		NfuIII	?	?	?	?	?	?	318
Neisseria gonorrhoea	G. Wilson	NgoI (HaeII)	PuGCGCPy	48	76	1	8	11	331
Neisseria gonorrhoea	CDC 66	NgoII (HaeIII)	GGCC	149	216	18	11	22	42
Neisseria gonorrhoea KH 7764-45	L. Mayer	NgoIII (SacII)	CCGCGG	4	33	0	1	0	214
Neisseria lactamica	NRCC 2118	NlaI (HaeIII)	GGCC	149	216	18	11	22	371
		NlaII (MboI)	GATC	116	87	8	0	22	371
		NlaIII	CATG↑	181	183	17	22	26	371
		NlaIV	GGN↑NCC	82	178	16	6	24	371
Neisseria lactamica	NRCC 31016	NlaSI (SacII)	CCGCGG	4	33	0	1	0	36
		NlaSII (AcyI)	GPuCGPyC	40	44	0	7	6	36
Neisseria meningitidis DRES-30	R. Sparling	NmeIV	?	?	?	?	?	?	282
Neisseria meningitidis DRES-W34	R. Sparling	NmeI	?	18	?	?	?	?	282
	R. Sparling	NmeII	?	28	?	0	0	?	282
Neisseria meningitidis M1011	R. Sparling	NmeIII	?	?	?	?	?	?	282
Neisseria mucosa	ATCC 25999	NheI	G↑CTAGC	1	4	0	0	1	48
Neisseria mucosa	ATCC 25997	NmuDI (DpnI)	GATC	cleaves methylated DNA					35
Neisseria mucosa	ATCC 25996	NmuEI (DpnI)	GATC	cleaves methylated DNA					30
		NmuEII (AsuI)	GGNCC	74	164	11	2	15	35
Neisseria mucosa	ATCC 19697	NmuFI (NaeI)	GCCGGC	1	13	1	0	4	35
Neisseria mucosa	NRCC 31013	NmuI (NaeI)	GCCGGC	1	13	1	0	4	318

Microorganism	Source	Enzyme	Sequence	λ	Ad2	SV40	ΦX	pBR	References
Moraxella osloensis	ATCC 19976	MnoIII (MboI)	GATC	116	87	8	0	22	211
Moraxella phenylpyruvica	ATCC 17955	MosI (MboI)	GATC	116	87	8	0	22	77
Moraxella species	R.J. Roberts	MphI (EcoRII)	CC(A/T)GG	71	136	17	2	6	138
Moraxella species MS67	M. Sargent	MspI (HpaII)	C↑CGG	328	171	1	5	26	312;312,250;137
		Msp67I (ScrFI)	CC↑NGG	185	233	17	3	16	368
		Msp67II (MboI)	GATC	116	87	8	0	22	368
Myxococcus stipitatus Mxs2	H. Reichenbach	MstI (XhoI)	CTCGAG	1	6	0	1	0	197,211
		MstII	?	?	?	?	?	?	190
Myxococcus virescens V-2	H. Reichenbach	MviI	?	1	?	?	?	?	204
		MviII	?	?	?	?	?	?	204
Neisseria animalis	ATCC 19573	NanI (EcoRV)	GATATC	21	9	1	0	1	360
		NanII (DpnI)	GATC	cleaves methylated DNA					360
Neisseria caviae	NRCC 31003	NcaI (HinfI)	GANTC	148	72	10	21	10	318
Neisseria cinerea	NRCC 31006	NciI (CauII)	CC↑(C/G)GG	114	97	0	1	10	323;112
Neisseria cuniculi	ATCC 14688	NcuI (MboII)	GAAGA	130	113	16	11	11	31
Neisseria denitrificans	NRCC 31009	NdeI	CA↑TATG	7	2	2	0	1	324
		NdeII (MboI)	GATC	116	87	8	0	22	318
Neisseria flavescens	ATCC 13120	NflAI (EcoRV)	GATATC	21	9	1	0	1	186
		NflAII (MboI)	GATC	116	87	8	0	22	186
Neisseria flavescens	NRCC 31011	NflI (MboI)	GATC	116	87	8	0	22	318
		NflII	?	?	?	?	?	?	318

Microorganism	Source	Enzyme	Sequence	Number of cleavage sites					References
				λ	Ad2	SV40	ΦX	pBR	
Neisseria mucosa	ATCC 19693	NmuSI (AsuI)	GGNCC	74	164	11	2	15	47
Neisseria ovis	NRCC 31020	NovI	?	?	?	?	?	?	318
		NovII (HinfI)	GANTC	148	72	10	21	10	318
Neisseria sicca	ATCC 9913	NsiAI (Sau3AI)	GATC	116	87	8	0	22	44
Neisseria sicca	NRCC 31004	NsiHI (HinfI)	GANTC	148	72	10	21	10	317
Neisseria sicca	ATCC 29256	NsiI (AvaIII)	ATGCA↑T	14	9	3	0	0	46
Neisseria subflava	ATCC 19243	NsuDI (DpnI)	GATC	cleaves methylated DNA					35
Neisseria subflava	ATCC 14221	NsuI (MboI)	GATC	116	87	8	0	22	35
Nocardia aerocolonigenes	ATCC 23870	NaeI	GCC↑GGC	1	13	1	0	4	51
Nocardia amarae	ATCC 27809	NamI (NarI)	GGCGCC	1	20	0	2	4	175
Nocardia argentinensis	ATCC 31306	NarI	GG↑CGCC	1	20	0	2	4	50
Nocardia asteroides	ATCC 9970	NasBI (BamHI)	GGATCC	5	3	1	0	1	361
Nocardia asteroides	ATCC 7372	NasI (PstI)	CTGCAG	28	30	2	1	1	359
Nocardia asteroides	ATCC 9969	NasSI (SacI)	GAGCTC	2	16	0	0	0	359
Nocardia blackwellii	ATCC 6846	NblI (PvuI)	CGAT↑CG	3	7	0	0	1	250
Nocardia brasiliensis	ATCC 19296	NbaI (NaeI)	GCCGGC	1	13	1	0	4	228
Nocardia brasiliensis	ATCC 27936	NbrI (NaeI)	GCCGGC	1	13	1	0	4	228
Nocardia corallina	ATCC 19070	NcoI	C↑CATGG	4	20	3	0	0	162
Nocardia dassonvillei	ATCC 21944	NdaI (NarI)	GG↑CGCC	1	20	0	2	4	45
Nocardia globerula	ATCC 21292	NgbI (PstI)	CTGCAG	28	30	2	1	1	361
Nocardia minima	ATCC 19150	NmiI (KpnI)	GGTACC	2	8	1	0	0	48

25

Microorganism	Source	Enzyme	Sequence	λ	Ad2	SV40	ΦX	pBR	References
Nocardia opaca	ATCC 21507	NopI (SalI)	G↑TCGAC	2	3	0	0	1	250
		NopII	?	?	?	?	?	?	162
Nocardia otitidis-caviarum	ATCC 14629	NocI (PstI)	CTGCAG	28	30	2	1	1	48
Nocardia otitidis-caviarum	ATCC 14630	NotI	GC↑GGCCGC	0	7	0	0	0	21;257
Nocardia rubra	ATCC 15906	NruI	TCG↑CGA	5	5	0	2	1	48
Nocardia species	ATCC 19170	NspAI (Sau3AI)	GATC	116	87	8	0	22	361
Nocardia tartaricans	ATCC 31191	NtaI (Tth111I)	GACNNNGTC	2	12	0	0	1	359
Nocardia tartaricans	ATCC 31190	NtaSI (StuI)	AGGCCT	6	11	7	1	0	359
		NtaSII (NaeI)	GCCGGC	1	13	1	0	4	359
Nocardia uniformis	ATCC 21806	NunI	?	?	?	?	?	?	162
		NunII (NarI)	GG↑CGCC	1	20	0	2	4	162
Nostoc species	PCC 7524	Nsp(7524)I	PuCATG↑Py	32	41	2	0	4	232
		Nsp(7524)II (SduI)	G(A/G/T)GC(A/C/T)↑C	38	105	4	3	10	232
		Nsp(7524)III (AvaI)	C↑PyCGPuG	8	40	0	1	1	232
		Nsp(7524)IV (AsuI)	G↑GNCC	74	164	11	2	15	232
		Nsp(7524)V (AsuII)	TTCGAA	7	1	0	0	0	232
Nostoc species	PCC 6705	NspBI (AsuII)	TTCGAA	7	1	0	0	0	63
		NspBII	C(A/C)G↑C(G/T)G	75	95	4	5	6	63
Nostoc species	PCC 7413	NspHI (Nsp(7524)I)	PuCATG↑Py	32	41	2	0	4	63
		NspHII (AvaII)	GG(A/T)CC	35	73	6	1	8	63
Nostoc species	PCC 8009	NspMACI (BglII)	A↑GATCT	6	11	0	0	0	164

Microorganism	Source	Enzyme	Sequence	Number of cleavage sites				References
				λ	Ad2	SV40	ΦX	pBR	
Nostoc species SA	D. Jones	NspSAI (AvaI)	C↑PyCGPuG	8	40	0	1	1	369
		NspSAII (BstEII)	G↑GTNACC	13	10	0	0	0	369
		NspSAIII (NcoI)	CCATGG	4	20	3	0	0	369
		NspSAIV (BamHI)	G↑GATCC	5	3	1	0	1	369
Oerskovia xanthineolytica	R. Shekman	OxaI (AluI)	AGCT	143	158	34	24	16	285
		OxaII	?	?	?	?	?	?	285
Oerskovia xanthineolytica N	NEB 401	OxaNI (SauI)	CCTNAGG	2	7	0	0	0	389
Proteus myxofaciens	ATCC 19692	PmyI (PstI)	CTGCAG	28	30	2	1	1	375
Proteus vulgaris	ATCC 13315	PvuI	CGAT↑CG	3	7	0	0	1	82
		PvuII	CAG↑CTG	15	24	3	0	1	82
Providencia alcalifaciens	ATCC 9886	PalI (HaeIII)	GGCC	149	216	18	11	22	78
Providencia stuartii 164	J. Davies	PstI	CTGCA↑G*	28	30	2	1	1	275;31;383
Pseudoanabaena species	ATCC 27263	PspI (AsuI)	G↑GNCC	74	164	11	2	15	206
Pseudomonas aeruginosa	N.N. Sokolov	PaeI (SphI)	GCATG↑C	6	8	2	0	1	281
Pseudomonas aeruginosa	G.A. Jacoby	PaeR7 (XhoI)	C↑TCGAG	1	5	0	0	0	107;81
Pseudomonas alkaligenes	ATCC 12815	PalI (HaeIII)	GGCC	149	216	18	11	22	290
Pseudomonas alkanolytica	IFO 12319	PanI (XhoI)	C↑TCGAG	1	6	0	1	0	290
Pseudomonas facilis	M. VanMontagu	PfaI (Sau3AI)	↑GATC	116	87	8	0	22	312
Pseudomonas fluorescens	IFO 3507	PfII	?	?	?	?	?	?	290
Pseudomonas fluorescens	NEB 375	PfIMI	CCANNNN↑NTGG	14	18	2	2	2	202
Pseudomonas fluorescens	T.S. Wang	PfIWI (XhoI)	CTCGAG	1	6	0	1	0	320

Microorganism	Source	Enzyme	Sequence	Number of cleavage sites				References
				λ	Ad2	SV40	ΦX	pBR	
Pseudomonas glycinae	J.V. Leary	PglI (NaeI)	GCCGGC	1	13	1	0	4	169
		PglII	?	0	>25	0	?	1	169
Pseudomonas maltophila	D. Comb	PmaI (PstI)	CTGCAG	28	30	2	1	1	250
Pseudomonas maltophila CB50P	C.A. Hart	PmaCI	CAC↑GTG	3	10	0	0	0	363
Pseudomonas mirabilis	N.L. Bakh	PmI	?	9	?	?	?	1	372
Pseudomonas ovis	S. Riazuddin	PovI (BclI)	TGATCA	8	5	1	0	0	378
Pseudomonas paucimobilis	NEB 376	PpaI	GAGACC	2	18	0	0	1	202
Pseudomonas putida C-83	Toyoboseki Co.	PpuI (HaeIII)	GGCC	149	216	18	11	22	290
Pseudomonas putida M	NEB 372	PpuMI	PuG↑G(A/T)CCPy	3	23	1	0	2	351;202
Pseudomonas species	R.A. Makula	PssI (DraII)	PuGGNC↑CPy	3	44	3	0	4	181,9
		PssII	?	?	?	?	?	?	181
Pseudomonas species MS61	M. Sargent	Psp6II (NarI)	GCCGGC	1	20	0	2	4	368
Rhizobium leguminosarum 300	J. Beringer	RleI	?	6	>10	?	?	?	333
Rhizobium lupini 1	W. Heumann	RluI (NaeI)	GCCGGC	1	13	1	0	4	332,106;274
Rhizobium meliloti	J.L. Denarie	RmeI	?	8	>10	?	?	?	106
Rhodococcus rhodochrous	ATCC 14349	RrhI (SalI)	GTCGAC	2	3	0	0	1	228
		RrhII	?	?	?	?	?	?	228
Rhodococcus rhodochrous	ATCC 4276	RroI (SalI)	GTCGAC	2	3	0	0	1	228
Rhodococcus species	ATCC 21664	RheI (SalI)	GTCGAC	2	3	0	0	1	116
Rhodococcus species	ATCC 19148	RhpI (SalI)	GTCGAC	2	3	0	0	1	116
		RhpII	?	2	?	?	?	?	116

Microorganism	Source	Enzyme	Sequence	λ	Ad2	SV40	ΦX	pBR	References
Rhodococcus species	ATCC 13259	RhsI (BamHI)	GGATCC	5	3	1	0	1	116
Rhodopseudomonas sphaeroides	S. Kaplan	RsaI	GT↑AC	113	83	12	11	3	180
Rhodopseudomonas sphaeroides		RshI (PvuI)	CGAT↑CG	3	7	0	0	1	179
Rhodopseudomonas sphaeroides	V.M. Kramarov	RshII (CauII)	CC(C/G)GG	114	97	0	1	10	155
Rhodopseudomonas sphaeroides	R. Lascelles	RspI (PvuI)	CGATCG	3	7	0	0	1	14
Rhodopseudomonas sphaeroides	S. Kaplan	RsrI (EcoRI)	GAATTC	5	5	1	0	1	73
		RsrII	CGtG(A/T)CCG	5	2	0	0	0	215
Rhodospirillum rubrum	J. Chirikjian	RrbI	?	?	4	5	1	?	171
Salmonella hybrid	L.R. Bullas	SQ	AACNNNNNTAPyG	Type I enzyme					353
Salmonella infantis	A. deWaard	SinI (AvaII)	GG(A/T)CC	35	73	6	1	8	178
Salmonella potsdam	L.R. Bullas	SP	AACNNNNNGTPuC*	Type I enzyme					352
Salmonella typhi 27	E.S. Anderson	StyI	C↑C(A/T)(A/T)GG*	10	44	8	0	1	199
Salmonella typhimurium	L.R. Bullas	SB	GAGNNNNNPuTAPyG*	Type I enzyme					352
Serratia fonticola	ATCC 29938	SfnI (AvaII)	GG(A/T)CC	35	73	6	1	8	359
Serratia fonticola	NEB 369	SfoI (NarI)	GGCGCC	1	20	0	2	4	202
Serratia marcescens Sb	C. Mulder	SmaI	CCC↑GGG	3	12	0	0	0	92;65
Serratia species SAI	B. Torheim	SspXI	?	?	?	?	?	?	305
Shigella boydii 13	NCTC 9361	Sbo13 (NruI)	TCGCGA	5	5	0	2	1	338
Shigella sonnei 47	T.M. Uporova	SsoI (EcoRI)	G↑AATTC	5	5	1	0	1	374
		SsoII (ScrFI)	↑CCNGG	185	233	17	3	16	374
Sphaerotilus natans	ATCC 15291	SnaBI	TAC↑GTA	1	0	0	0	0	20

Microorganism	Source	Enzyme	Sequence	Number of cleavage sites				References
				λ	Ad2	SV40	ΦX	pBR	
Sphaerotilus natans	ATCC 13923	SpeI	A↑CTAGT	0	3	0	0	0	48
Sphaerotilus natans	ATCC 13925	SspI	AAT↑ATT	20	5	1	6	1	89
Sphaerotilus natans C	A. Pope	SnaI	GTATAC	3	3	0	0	1	225
Spiroplasma citri ASP2	M.A. Stephens	SciNI (HhaI)	G↑CGC	215	375	2	18	31	283
Spirulina platensis	M. Kawamura	SpII	C↑TGTACG	1	4	0	2	0	370
		SpII (Tth111I)	GACNNNGTC	2	12	0	0	1	370
		SpIII (HaeIII)	GGCC	149	216	18	11	22	370
Staphylococcus aureus 3A	E.E. Stobberingh	Sau3AI (MboI)	↑GATC	116	87	8	0	22	294
Staphylococcus aureus PS96	E.E. Stobberingh	Sau96I (AsuI)	G↑GNCC	74	164	11	2	15	295
Staphylococcus intermedius	ATCC 29663	SinMI (MboI)	GATC	116	87	8	0	22	35
		SinMII	?	?	?	?	?	?	35
Staphylococcus saprophyticus	ATCC 13518	SsaI	?	>10	?	?	?	?	162
Streptococcus cremoris F	C. Daly	ScrFI	CC↑NGG	185	233	17	3	16	69
Streptococcus durans	A. Janulaitis	SduI	G(A/G/T)GC(A/C/T)↑C	38	105	4	3	10	126
Streptococcus dysgalactiae	ATCC 9926	SdyI (AsuI)	GGNCC	74	164	11	2	15	228
Streptococcus faecalis GU	J. Chirikjian	SfaGUI (HpaII)	CCGG	328	171	1	5	26	43
Streptococcus faecalis ND547	D. Clewell	SfaNI	GCATC (5/9)	169	84	6	12	22	260
Streptococcus faecalis var. zymogenes	R. Wu	SfaI (HaeIII)	GG↑CC	149	216	18	11	22	334
Streptomyces achromogenes	ATCC 12767	SacI	GAGCT↑C	2	16	0	0	0	2
		SacII	CCGC↑GG	4	33	0	1	0	2

Microorganism	Source	Enzyme	Sequence	Number of cleavage sites				References
				λ	Ad2	SV40	ΦX	pBR	
				100	100	?	?	?	2
Streptomyces alanosinicus	ATCC 15710	SacI (SacII)	CCGCGG	4	33	0	1	0	202
Streptomyces albofaciens	ATCC 25184	SaoI (NaeI)	GCCGGC	1	13	1	0	4	390
Streptomyces albohelvatus	ATCC 19820	SabI (SacII)	CCGCGG	4	33	0	1	0	390
Streptomyces albus	CMI 52766	SalPI (PstI)	CTGCA↑G	28	30	2	1	1	39;37
Streptomyces albus G	J.M. Ghuysen	SaII	G↑TCGAC	2	3	0	0	1	3
		SaII	?	>20	?	?	?	?	3
Streptomyces albus subspecies pathocidicus	KCC S-0166	SpaI (XhoI)	CTCGAG	1	6	0	1	0	270
Streptomyces aureofaciens	CCM 3239	Sau3239I	?	1	?	?	?	?	75
Streptomyces aureofaciens IKA 18/4	J. Timko	SauI	CC↑TNAGG	2	7	0	0	0	302
Streptomyces bobili	ATCC 3310	SboI (SacII)	CCGCGG	4	33	0	1	0	270;299
Streptomyces caespitosus	H. Takahashi	ScaI	AGT↑ACT	5	5	0	0	1	152;89
Streptomyces coelicolor	ATCC 10147	ScoI (SacI)	GAGCTC	2	16	0	0	0	390
Streptomyces cupidosporus	KCC S0316	ScuI (XhoI)	CTCGAG	1	6	0	1	0	270
Streptomyces exfoliatus	KCC S0030	SexI (XhoI)	CTCGAG	1	6	0	1	0	270
		SexII	?	2	?	?	?	?	270
Streptomyces fimbriatus	ATCC 15051	SfiI	GGCCNNNN↑NGGCC	0	3	1	0	0	229
Streptomyces fradiae	ATCC 3355	SfrI (SacII)	CCGCGG	4	33	0	1	0	270;299
Streptomyces ganmycicus	KCC S0759	SgaI (XhoI)	CTCGAG	1	6	0	1	0	270
Streptomyces goshikiensis	KCC S0294	SgoI (XhoI)	CTCGAG	1	6	0	1	0	270

Microorganism	Source	Enzyme	Sequence	Number of cleavage sites					References
				λ	Ad2	SV40	ΦX	pBR	
Streptomyces griseus	ATCC 23345	SgrI	?	0	7	0	?	?	2
Streptomyces griseus Kr. 20	A.V. Orekhov	SgrII (EcoRII)	CC(A/T)GG	71	136	17	2	6	218
Streptomyces hygroscopicus	F. Walter	ShyI (SacII)	CCGCGG	4	33	0	1	0	319
Streptomyces hygroscopicus	T. Yamaguchi	ShyTI	?	2	?	?	?	?	270
Streptomyces karnatakensis	ATCC 25463	SkaI (NaeI)	GCCGGC	1	13	1	0	4	35
		SkaII (PstI)	CTGCAG	28	30	2	1	1	35
Streptomyces lavendulae	ATCC 8664	SlaI (XhoI)	C↑TCGAG	1	6	0	1	0	300
Streptomyces luteoreticuli	KCC S0788	SluI (XhoI)	CTCGAG	1	6	0	1	0	299
Streptomyces novocastria	P. Eastlake	SnoI (ApaLI)	G↑TGCAC	4	7	0	1	3	366,27
Streptomyces oderifer	ATCC 6246	SodI	?	?	?	?	?	?	162
		SodII	?	?	?	?	?	?	162
Streptomyces phaeochromogenes	IFO 3108	SpaXI (SphI)	GCATGC	6	8	2	0	1	290
Streptomyces phaeochromogenes	F. Bolivar	SphI	GCATG↑C	6	8	2	0	1	72
Streptomyces stanford	S. Goff	SstI (SacI)	GAGCT↑C	2	16	0	0	0	86,205
		SstII (SacII)	CCGC↑GG	4	33	0	1	0	86
		SstIII (SacIII)	?	100	100	?	?	?	86
		SstIV (BclI)	TGATCA	8	5	1	0	0	111
Streptomyces tubercidicus	H. Takahashi	StuI	AGG↑CCT	6	11	7	1	0	269
Streptoverticillium cinnamonium	S.T. Williams	ScII (XhoI)	CTC↑GAG	1	6	0	1	0	368
Streptoverticillium flavopersicum	Upjohn UC 5066	SfII (PstI)	CTGCA↑G	28	30	2	1	1	138

Microorganism	Source	Enzyme	Sequence	Number of cleavage sites				References
				λ	Ad2	SV40	ΦX	pBR	
Sulfolobus acidocaldarius	DSM 639	SuaI (HaeIII)	GG↑CC	149	216	18	11	22	377
Sulfolobus acidocaldarius	W. Zillig	SuiI (HaeIII)	GGCC	149	216	18	11	22	193
Synechocystis species 6701	ATCC 27170	SecI	C↑CNNGG	105	234	16	6	8	345
		SecII (MspI)	CCGG	0	0	0	0	0	345
		SecIII (MstII)	CCTNAGG	2	6	0	0	0	345
Thermococcus celer	W. Zillig	TceI (MboII)	GAAGA	130	113	16	11	11	193
Thermoplasma acidophilum	D. Searcy	ThaI (FnuDII)	CG↑CG	157	303	0	14	23	191
Thermopolyspora glauca	ATCC 15345	TglA (SacII)	CCGCGG	4	33	0	1	0	85
Thermus aquaticus	S.A. Grachev	TaqXI (BstNI)	CC↑(A/T)GG	71	136	17	2	6	87
Thermus aquaticus YTI	J.I. Harris	TaqI	T↑CGA*	121	50	1	10	7	247;247;248
		TaqII	GACCGA (11/9) CACCCA (11/9)	28	36	1	2	6	211,6
Thermus flavus AT62	T. Oshima	TflI (TaqI)	TCGA	121	50	1	10	7	248
Thermus thermophilus HB8	T. Oshima	TthHB8I (TaqI)	TCGA*	121	50	1	10	7	249,315;249,315;248;248
Thermus thermophilus strain 110	T. Oshima	TteI (Tth111I)	GACNNNGTC	2	12	0	0	1	273
Thermus thermophilus strain 111	T. Oshima	Tth111I	GACN↑NNGTC	2	12	0	0	1	273
		Tth111II	CAAPuCA (11/9)	49	53	11	11	5	272
		Tth111III	?	?	?	?	?	?	271
Thermus thermophilus strain 23	T. Oshima	TtrI (Tth111I)	GACNNNGTC	2	12	0	0	1	273
Tolypothrix tenuis	W. Siegelman	TtnI (HaeIII)	GGCC	149	216	18	11	22	286

Microorganism	Source	Enzyme	Sequence	λ	Ad2	SV40	ΦX	pBR	References
Vibrio harveyi	ATCC 14126	VhaI (HaeIII)	GGCC	149	216	18	11	22	116
Xanthomonas amaranthicola	ATCC 11645	XamI (SalI)	GTCGAC	2	3	0	0	1	3
Xanthomonas badrii	ATCC 11672	XbaI	T↑CTAGA	1	5	0	0	0	343
Xanthomonas citrii	IFO 3835	XciI (SalI)	G↑TCGAC	2	3	0	0	1	325
Xanthomonas cyanopsidis 13D5	C.I. Kado	XcyI (SmaI)	C↑CCGGG	3	12	0	0	0	70
Xanthomonas holcicola	ATCC 13461	XhoI	C↑TCGAG	1	6	0	1	0	84
		XhoII	Pu↑GATCPy	21	22	3	0	8	217;85,154
Xanthomonas malvacearum	ATCC 9924	XmaI (SmaI)	C↑CCGGG	3	12	0	0	0	65
		XmaII (PstI)	CTGCAG	28	30	2	1	1	65
		XmaIII	C↑GGCCG	2	19	0	0	1	158
Xanthomonas manihotis 7AS1	B-C. Lin	XmnI	GAANN↑NNTTC	24	5	0	3	2	174,228
Xanthomonas nigromaculans	ATCC 23390	XnII (PvuI)	CGATCG	3	7	0	0	1	99
Xanthomonas oryzae	M. Ehrlich	XorI (PstI)	CTGCAG	28	30	2	1	1	321
		XorII (PvuI)	CGAT↑CG	3	7	0	0	1	321;82
Xanthomonas papavericola	ATCC 14180	XpaI (XhoI)	C↑TCGAG	1	6	0	1	0	84
Xanthomonas phaseoli	Z.F. Bunina	XphI (PstI)	CTGCAG	28	30	2	1	1	32
Zymomonas anaerobia	O.J. Yoo	ZanI (EcoRII)	CC(A/T)↑GG	71	136	17	2	6	292

REFERENCES
1. Agarwal, K. unpublished observations.
2. Arrand, J.R., Myers, P.A. and Roberts, R.J. unpublished observations.
3. Arrand, J.R., Myers, P.A. and Roberts, R.J. (1978) J. Mol. Biol. 118: 127-135.
4. Azizbekyan, R.R., Rebentish, B.A., Stepanova, T.V., Netyksa, E.M. and Buchkova, M.A. (1984) Dokl. Akad. Nauk. SSSR 274: 742-744.
5. Bachi, B., Reiser, J. and Pirrotta, V. (1979) J. Mol. Biol. 128: 143-163.
6. Barker, D., Hoff, M., Oliphant, A. and White, R. (1984) Nucl. Acids Res. 12: 5567-5581.
7. Baumstark, B.R., Roberts, R.J. and RajBhandray, U.L. (1979) J. Biol. Chem. 254: 8943-8950.
8. Beaty, J.S., McLean-Bowen, C.A. and Brown, L.R. (1982) Gene 18: 61-67.
9. Belle Isle, H. unpublished observations.
10. Bennett, S.P. and Halford, S.E. unpublished observations.
11. Bickle, T.A. and Ineichen, K. (1980) Gene 9: 205-212.
12. Bickle, T., Yuan, R., Pirrotta, V. and Ineichen, K. unpublished observations.
13. Bigger, C.H., Murray, K. and Murray, N.E. (1973) Nature New Biology 244: 7-10.
14. Bingham, A.H.A., Atkinson, A. and Darbyshire, J. unpublished observations.
15. Bingham, A.H.A., Atkinson, T., Sciaky, D. and Roberts, R.J. (1978) Nucl. Acids Res. 5: 3457-3467.
16. Bingham, A.H.A. and Darbyshire, J. (1982) Gene 18: 87-91.
17. Bingham, A.H.A., Sharp, R.J. and Atkinson, T. unpublished observations.
18. Bitinaite, J.B., Klimasauskas, S.J., Buktus, V.V. and Janulaitis, A.A. (1985) FEBS Letters 182: 509-513.
19. Bolton, B., Nesch, G., Comer, M., Wolf, W. and Kessler, C. (1985) FEBS Letters 182: 130-134.
20. Borsetti, R., Grandoni, R. and Schildkraut, I. unpublished observations.
21. Borsetti, R., Wise, D. and Schildkraut, I. unpublished observations.
22. Boyer, H.W., Chow, L.T., Dugaiczyk, A., Hedgpeth, J. and Goodman, H.M. (1973) Nature New Biology 224: 40-43.
23. Brockes, J.P. (1973) Biochem J. 133: 629-633.
24. Brockes, J.P., Brown, P.R. and Murray, K. (1972) Biochem. J. 127: 1-10.
25. Bron, S. and Murray, K. (1975) Mol. Gen. Genet. 143: 25-33.
26. Bron, S., Murray, K. and Trautner, T.A. (1975) Mol Gen. Genet. 143: 13-23.

27. Brown, N.L. unpublished observations.
28. Brown, N.L. Hutchinson, C.A. III and Smith, M. (1980) J. Mol. Biol. 140: 143-148.
29. Brown, N.L., McClelland, M. and Whitehead, P.R. (1980) Gene 9: 49-68.
30. Brown, N.L. and Smith, M. (1977) Proc. Natl. Acad. Sci. USA 74: 3213-3216.
31. Brown, N.L. and Smith, M. (1976) FEBS Letters 65: 284-287.
32. Bunina, Z.F., Kramarov, V.M., Smolyaninov, V.V. and Tolstova, L.A. (1984) Bioorg. Khim. 10: 1333-1335.
33. Butkus, V., Klimasauskas, S., Kersulyte, D., Vaitkevicius, D., Lebionka, A. and Janulaitis, A. (1985) Nucl. Acids Res. 13: 5727-5746.
34. Calleja, F., Dekker, B.M.M., Coursin, T. and deWaard, A. (1984) FEBS Letters 178: 69-72.
35. Camp, R. and Schildkraut, I. unpublished observations.
36. Camp, R. and Visentin, L.P. unpublished observations.
37. Carter, J.A., Chater, K.F., Bruton, C.J. and Brown, N.L. (1980) Nucl. Acids Res. 8: 4943-4954.
38. Catterall, J.F. and Welker, N.E. (1977) J. Bacteriol. 129: 1110-1120.
39. Chater, K.F. (1977) Nucl. Acids Res. 4: 1989-1998.
40. Chirikjian, J.G., George, A. and Smith, L.A. (1978) Fed. Proc. 37: 1415.
41. Clarke, C.M. and Hartley, B.S. (1979) Biochem. J. 177: 49-62.
42. Clanton, D.J., Woodward, J.M. and Miller, R.V. (1978) J. Bacteriol. 135: 270-273.
43. Coll, E. and Chirikjian, J. unpublished observations.
44. Comb, D.G. unpublished observations.
45. Comb, D.G., Hess, E.J. and Wilson, G. unpublished observations.
46. Comb, D.G., Parker, P., Grandoni, R. and Schildkraut, I. unpublished observations.
47. Comb, D.G., Parker, P. and Schildkraut, I. unpublished observations.
48. Comb, D.G. and Schildkraut, I. unpublished observations.
49. Comb, D.G., Schildkraut, I. and Roberts, R.J. unpublished observations.
50. Comb, D.G., Schildkraut, I., Wilson, G. and Greenough, L. unpublished observations.
51. Comb, D.G. and Wilson, G. unpublished observations.
52. Cruz, A.K., Kidane, G., Pires, M.Q., Rabinovitch, L., Guaycurus, T.V. and Morel, C.M. (1984) FEBS Letters 173: 99-102.
53. de Waard, A. unpublished observations.
54. de Waard, A. and Duyvesteyn, M. (1980) Arch. Microbiol. 128: 242-247.
55. de Waard, A., Korsuize, J., van Beveren, C.P. and Maat, J. (1978) FEBS Letters 96: 106-110.

56. de Waard, A., van Beveren, C.P., Duyvesteyn, M. and van Ormondt, H. (1979) FEBS Letters 101: 71-76.
57. de Wit, C.M., Dekker, B.M.M., Neele, A.C. and de Waard, A. (1985) FEBS Letters 180: 219-223.
58. DiLauro, R. unpublished observations.
59. Dugaiczyk, A., Hedgpeth, J., Boyer, H.W. and Goodman, H.M. (1974) Biochemistry 13: 503-512.
60. Duncan, C.H., Wilson, G.A. and Young, F.E. (1978) J. Bacteriol. 134: 338-344.
61. Duyvesteyn, M.G.C. and de Waard, A. (1980) FEBS Letters 111: 423-426.
62. Duyvesteyn, M.G.C., Korsuize, J. and de Waard, A. (1981) Plant Mol. Biol. 1: 75-79.
63. Duyvesteyn, M.G.C., Korsuize, J., de Waard, A., Vonshak, A. and Wolk, C.P. (1983) Arch. Microbiol. 134: 276-281.
64. Endow, S.A. (1977) J. Mol. Biol. 114: 441-449.
65. Endow, S.A. and Roberts, R.J. (1977) J. Mol. Biol. 112: 521-529.
66. Endow, S.A. and Roberts, R.J. unpublished observations.
67. Eskin, B. and Linn, S. (1972) J. Biol. Chem. 247: 6183-6191.
68. Fisherman, J., Gingeras, T.R. and Roberts, R.J. unpublished observations.
69. Fitzgerald, G.F., Daly, C., Brown, L.R. and Gingeras, T.R. (1982) Nucl. Acids Res. 10: 8171-8179.
70. Froman, B.E., Tait, R.C., Kado, C.I. and Rodriguez, R.L. (1984) Gene 28: 331-335.
71. Fuchs, C., Rosenvold, E.C., Honigman, A. and Szybalski, W. (1978) Gene 4: 1-23.
72. Fuchs, L.Y., Covarrubias, L., Escalante, L., Sanchez, S. and Bolivar, F. (1980) Gene 10 39-46.
73. Gardner, J.F., Cohen, L.K., Lynn, S.P. and Kaplan, S. unpublished observations.
74. Garfin, D.E. and Goodman, H.M. (1974) Biochem. Biophys. Res. Comm. 59: 108-116.
75. Gasperik, J., Godany, A., Hostinova, E. and Zelinka, J. (1983) Biologia (Bratislava) 38: 315-319.
76. Geier, G.E. and Motrich, P. (1979) J. Biol. Chem. 254: 1408-1413.
77. Gelinas, R.E., Myers, P.A. and Roberts, R.J. (1977) J. Mol. Biol. 114: 169-179.
78. Gelinas, R.E., Myers, P.A. and Roberts, R.J. unpublished observations.
79. Gelinas, R.E., Myers, P.A., Weiss, G.A., Murray, K. and Roberts, R.J. (1977) J. Mol. Biol. 114: 433-440.
80. Gelinas, R.E. and Roberts, R.J. unpublished observations.

81. Gingeras, T.R. and Brooks, J.E. (1983) Proc. Natl. Acad. Sci. USA $\underline{80}$: 402-406.
82. Gingeras, T.R., Greenough, L., Schildkraut, I. and Roberts, R.J. (1981) Nucl. Acids Res. $\underline{9}$: 4525-4536.
83. Gingeras, T.R., Milazzo, J.P. and Roberts, R.J. (1979) Nucl. Acids Res. $\underline{5}$: 4105-4127.
84. Gingeras, T.R., Myers, P.A., Olson, J.A., Hanberg, F.A. and Roberts, R.J. (1978) J. Mol. Biol. $\underline{118}$: 113-122.
85. Gingeras, T.R. and Roberts, R.J. unpublished observations.
86. Goff, S.P. and Rambach, A. (1978) Gene $\underline{3}$: 347-352.
87. Grachev, S.A., Mamaev, S.V. Gurevich, A.I., Igoshin, A.V., Kolosov, M.N. and Slyusarenko, A.G. (1981) Bioorg. Khim. $\underline{7}$: 628-630.
88. Grandoni, R.P. and Comb, D. unpublished observations.
89. Grandoni, R.P. and Schildkraut, I. unpublished observations.
90. Greenaway, P.J. (1980) Biochem. Biophys. Res. Comm. $\underline{95}$: 1282-1287.
91. Greene, P.J., Betlach, M.C., Boyer, H.W. and Goodman, H.M. (1974) Methods Mol. Biol. $\underline{7}$: 87-111.
92. Greene, R. and Mulder, C. unpublished observations.
93. Grosskopf, R., Wolf, W. and Kessler, C. (1985) Nucl. Acids Res. $\underline{13}$: 1517-1528.
94. Grosveld, G.C. unpublished observations.
95. Gunthert, U., Storm, K. and Bald, R. (1978) Eur. J. Biochem. $\underline{90}$: 581-583.
96. Haberman, A. (1974) J. Mol. Biol. $\underline{89}$: 545-563.
97. Haberman, A., Heywood, J. and Meselson, M. (1972) Proc. Natl. Acad. Sci. USA $\underline{69}$: 3138-3141.
98. Hadi, S.M., Bachi, B., Shepherd, J.C.W., Yuan, R., Ineichen, K. and Bickle, T.A. (1979) J. Mol. Biol. $\underline{134}$: 655-666.
99. Hanberg, F., Myers, P.A. and Roberts, R.J. unpublished observations.
100. Hansen, R. unpublished observations.
101. Hartmann, H. and Goebel, W. (1977) FEBS Letters $\underline{80}$: 285-287.
102. Hattman, S., Brooks, J.E. and Masurekar, M. (1978) J. Mol. Biol. $\underline{126}$: 367-380.
103. Hattman, S., Keisler, T. and Gottehrer, A. (1978) J. Mol. Biol. $\underline{124}$: 701-711.
104. Hedgpeth, J., Goodman, H.M. and Boyer, H.W. (1972) Proc. Natl. Acad. Sci. USA $\underline{69}$: 3448-3452.
105. Hendrix, J.D. and Welker, N.E. (1985) J. Bacteriol. $\underline{162}$: 682-692.
106. Heumann, W. (1979) Curr. Top. Microbiol. Immunol. $\underline{88}$: 1-24.
107. Hinkle, N.F. and Miller, R.V. (1979) Plasmid $\underline{2}$: 387-393.
108. Hiraoka, N., Kita, K., Nakajima, H. and Obayashi, A. (1984) J. Ferment.

Technol. 62: 583-588.
109. Hobom, G., Schwarz, E., Melzer, M. and Mayer, H. (1981) Nucl. Acids Res. 9: 4823-4832.
110. Hoshino, T., Uozumi, T., Horinouchi, S., Ozaki, A., Beppu, T. and Arima, K. (1977) Biochim. Biophys. Acta 479: 367-369.
111. Hu, A.W., Kuebbing, D. and Blakesley, R.J. (1978) Fed. Proc. 38: 780.
112. Hu, A.W. and Marschel, A.H. unpublished observations.
113. Hughes, S.G., Bruce, T. and Murray, K. (1980) Biochem. J. 185: 59-63.
114. Hughes, S.G., Bruce, T. and Murray, K. unpublished observations.
115. Hughes, S.G. and Murray, K. (1980) Biochem. J. 185: 65-75.
116. Hurlin, P. and Schildkraut, I. unpublished observations.
117. Hutchinson, C.A. and Barrell, B.G. unpublished observations.
118. Ikawa, S., Shibata, T. and Ando, T. (1976) J. Biochem. (Tokyo) 80: 1457-1460.
119. Ikawa, S., Shibata, T. and Ando. T. (1979) Agric. Biol. Chem. 43: 873-875.
120. Ikawa, S., Shibata, T., Ando, T. and Saito, H. (1980) Molec. Gen. Genet. 177: 359-368.
121. Janulaitis, A. and Adomaviciute, L. unpublished observations.
122. Janulaitis, A. and Bitinaite, J. unpublished observations.
123. Janulaitis, A., Bitinaite, J. and Jaskeleviciene, B., (1983) FEBS Letters 151: 234-247.
124. Janulaitis, A., Klimasauskas, S., Petrusyte, M. and Butkus, V. (1983) FEBS Letters 161: 131-134.
125. Janulaitis, A., Marcinkeviciene, L.U. and Petrusyte, M.P. (1982) Dokl. Akad. Nauk. SSSR 577: 241-244.
126. Janulaitis, A., Marcinkeviciene, L., Petrusyte, M. and Mironov, A. (1981) FEBS Letters 134: 172-174.
127. Janulaitis, A. and Petrusyte, M. unpublished observations.
128. Janulaitis, A., Petrusyte, M. and Buktus, V.V. FEBS Letters, in press.
129. Janulaitis, A., Petrusyte, M., and Jaskelavicene, B.P., Krayev, A.S., Skryabin, K.G. and Bayev, A.A. (1981) Dokl. Akad. Nauk. SSSR 257: 749-750.
130. Janulaitis, A., Petrusite, M., and Jaskelavicene, B.P., Krayev, A.S., Skryabin, K.G. and Bayev, A.A. (1982) FEBS Letters 137: 178-180.
131. Janulaitis, A., Stakenas, P.S. and Berlin, Y. (1983) FEBS Letters 161: 210-212.
132. Janulaitis, A., Stakenas, P.S., Bitinaite, J.B. and Jaskeleviciene, B.P. (1983) Dolk. Akad. Nauk. SSSR 271: 483-485.

133. Janulaitis, A., Stakenas, P.S., Jaskeleviciene, B.P., Lebedenko, E.N. and Berlin, Y.A. (1980) Bioorg. Khim 6: 1746-1748.
134. Janulaitis, A., Stakenas, P.S., Petrusyte, M.P., Bitinaite, Yu.B., Klimashauskas, S.I. and Buktus, V.V. (1983) Molekulyarnaya Biologiya 18: 115-129.
135. Janulaitis, A.A., Vaitkevicius, D., Puntezis, S. and Jaskeleviciene, B. unpublished observations.
136. Jentsch, S. (1983) J. Bacteriol. 156: 800-808.
137. Jentsch, S., Gunthert, U. and Trautner, T.A. (1981) Nucl. Acids Res. 12: 2753-2759.
138. Jiang, B.D. and Myers, P. unpublished observations.
139. Kan, N.C., Lautenberger, J.A., Edgell, M.H. and Hutchinson, C.A. III (1979) J. Mol. Biol. 130: 191-209.
140. Kauc, L. and Piekarowicz, A. (1978) Eur. J. Biochem. 92: 417-426.
141. Kazennova, E.V. and Tarasov, A.P., Mileikovskaya, M.M., Semina, I.E. and Tsvetkova, N.V. (1982) Zh. Mikrobiol. Epidemiol. Immunobiol. 0: 56-57.
142. Kelly, T.J., Jr. and Smith, H.O. (1970) J. Mol. Biol. 51: 393-409.
143. Kessler, C. Neumaier, P.S. and Wolf, W. (1985) Gene 33: 1-102.
144. Kholmina, G.V., Rebentish, B.A., Skoblov, Y.S., Mironov, A.A., Yankovsky, N.K., Kozlov, Y.I., Glatman, L.I., Moroz, A.F. and Debabov, V.G. (1980) Dokl. Akad. Nauk. SSSR 253: 495-497.
145. Khosaka, T. unpublished observations.
146. Khosaka, T. and Kiwaki, M. (1984) Gene 31: 251-255.
147. Khosaka, T. and Kiwaki, M. (1984) FEBS Letters 177: 57-60.
148. Khosaka, T., Sakaurai, T., Takahashi, H. and Saito, H. (1982) Gene 17: 117-122.
149. Kiss, A., Sain, B., Csordas-Toth, E. and Venetianer, P. (1977) Gene 1: 323-329.
150. Kita, K., Hiroaka, N., Kimizuka, F. and Obayashi, A. (1984) Agric. Biol. Chem. 48: 531-532.
151. Kleid, D., Humayun, Z., Jeffrey, A. and Ptashne, M. (1976) Proc. Natl. Acad. Sci. USA 73: 293-297.
152. Kojima, H. Takahashi, H. and Saito, H. unpublished observations.
153. Koncz, C., Kiss, A. and Venetianer, P. (1978) Eur. J. Biochem. 89: 523-529.
154. Kramarov, V.M., Mazanov, A.L. and Smolyaninov, V.V. (1982) Bioorg. Khim. 8: 220-223.
155. Kramarov, V.M. Pachkunov, D.M. and Matvienko, N.I. (1983) in Nek. Aspekty Fiziol. Mikroog., Akad. Nauk SSSR (ed. Gaziev, A.I.), Nauchn. Tsentr Biol. Issled., Puschino, USSR, pp 22-26.

156. Kroger, M. and Hobom, G. (1984) Nucl. Acids Res. 12: 887-899.
157. Kroger, M., Hobom, G., Schutte, H. and Mayer, H. (1984) Nucl. Acids Res. 12: 3127-3141.
158. Kunkel, L.M., Silberklang, M. and McCarthy, B.J. (1979) J. Mol. Biol. 132: 133-139.
159. Lacks, S. and Greenberg, B. (1975) J. Biol. Chem. 250: 4060-4066.
160. Lacks, S. and Greenberg, B. (1977) J. Mol. Biol. 114: 153-168.
161. Landy, A., Ruedisueli, E., Robinson, L., Foeller, C. and Ross, W. (1974) Biochemistry 13: 2134-2142.
162. Langdale, J.A., Myers, P.A. and Roberts, R.J. unpublished observations.
163. Lau, R.H. and Doolittle, W.F. (1980) FEBS Letters 12: 200-202.
164. Lau, R.H., Visentin, L.P., Martin, S.M., Hofman, J.D. and Doolittle, W.F. (1985) FEBS Letters 179: 129-132.
165. Lautenberger, J.A., Edgell, M.H. and Hutchinson, C.A. III (1980) Gene 12: 171-174.
166. Lautenberger, J.A., Kan, N.C., Lackey, D., Linn, S., Edgell, M.H. and Hutchinson, C.A. III (1978) Proc. Natl. Acad. Sci. USA 75: 2271-2275.
167. Lautenberger, J.A. and Linn, S. (1972) J. Biol. Chem. 247: 6176-6182.
168. Lautenberger, J.A., White, C.T., Haigwood, N.L., Edgell, M.H. and Hutchinson, C.A. III (1980) Gene 9: 213-231.
169. Leary, J.V. unpublished observations.
170. LeBon, J.M., Kado, C., Rosenthal, L.J. and Chirikjian, J. (1978) Proc. Natl. Acad. Sci. USA 75: 4097-4101.
171. LeBon, J.M., LeBon, T., Blakesley, R. and Chirikjian, J. unpublished observations.
172. Leung, D.W., Lui, A.C.P., Merilees, H., McBride, B.C. and Smith, M. (1979) Nucl. Acids Res. 6: 17-25.
173. Levi, C. and Bickle, T. unpublished observations.
174. Lin, B-C., Chien, M-C. and Lou, S-Y. (1980) Nucl. Acids Res. 8: 6189-6198.
175. Lin, P-M. and Roberts, R.J. unpublished observations.
176. Lui, A.C.P., McBride, B.C. and Smith, M. unpublished observations.
177. Lui, A.C.P., McBride, B.C., Vovis, G.F. and Smith, M. (1979) Nucl. Acids Res. 6: 1-15.
178. Lupker, H.S.C. and Dekker, B.M.M. (1981) Biochim. Biophys. Acta 654: 297-299.
179. Lynn, S.P., Cohen, L.K., Gardner, J.F. and Kaplan, S. (1979) J. Bacteriol. 138: 505-509.
180. Lynn, S.P., Cohen, L.K., Kaplan, S. and Gardner, J.F. (1980) J. Bacteriol. 142: 380-383.
181. Makula, R.A. unpublished observations.

182. Makula, R.A. and Meagher, R.B. (1980) Nucl. Acids Res. 8: 3125-3131.
183. Mann, M.B., Rao, R.N. and Smith, H.O. (1978) Gene 3: 97-112.
184. Mann, M.B. and Smith, H.O. (1977) Nucl. Acids Res. 4: 4211-4221.
185. Mann, M.B. and Smith, H.O. unpublished observations.
186. Maratea, E. and Camp, R.R. unpublished observations.
187. Matvienko, N.I., Pachkunov, D.M. and Kramarov, V.M. (1984) FEBS Letters 177: 23-26.
188. Mayer, H., Grosschedl, R., Schutte, H. and Hobom, G. (1981) Nucl. Acids Res. 9: 4833-4845.
189. Mayer, H. and Klaar, J. unpublished observations.
190. Mayer, H. and Schutte, H. unpublished observations.
191. McConnell, D.J., Searcy, D.G. and Sutcliffe, J.G. (1978) Nucl. Acids Res. 5: 1729-1739.
192. McEvoy, S. and Roberts, R.J. unpublished observations.
193. McWilliam, P. quoted in reference 143.
194. Meagher, R.B. unpublished observations.
195. Meselson, M. and Yuan, R. (1968) Nature 217: 1110-1114.
196. Middleton, J.H. Edgell, M.H. and Hutchison, C.A. III (1972) J. Virol. 10: 42-50.
197. Middleton, J.H., Stankus, P.V., Edgell, M.H. and Hutchinson, C.A. III unpublished observations.
198. Mise, K. and Nakajima, K. (1984) Gene 30: 79-85.
199. Mise, K. and Nakajima, K. (1985) Gene 33: 357-361.
200. Mise, K. and Nakajima, K. (1985) Gene 36: 363-367.
201. Molemans, F., van Emmelo, J. and Fiers, W. (1982) Gene 18: 93-96.
202. Morgan, R. unpublished observations.
203. Morgan, R. and Hoffman, L. unpublished observations.
204. Morris, D.W. and Parish, J.H. (1976) Arch. Microbiol. 108: 227-230.
205. Muller, F., Stoffel, S. and Clarkson, S.G. unpublished observations.
206. Mulligan, B.J. and Szekeres, M. unpublished observations.
207. Murray, K., Brown, J.S. and Bruce, S.A. unpublished observations.
208. Murray, K. Hughes, S.G., Brown, J.S. and Bruce, S.A. (1976) Biochem. J. 159: 317-322.
209. Murray, K. and Morrison, A. unpublished observations.
210. Murray, K., Morrison, A., Cooke, H.W. and Roberts, R.J. unpublished observations.
211. Myers, P.A. and Roberts, R.J. unpublished observations.
212. Nagaraja, V., Stieger, M., Nager, C., Hadi, S.M. and Bickle, T. (1985) Nucl. Acids Res. 13: 389-399.
213. Nardone, G. and Blakesley, G. (1981) Fed. Proc. 40: 1848.

214. Norlander, L., Davies, J.K., Hagblom, P., and Normark, S. (1981) J. Bacteriol. 145: 788-795.
215. O'Connor, C.D., Metcalf, E., Wrighton, C.J., Harris, T.J.R. and Saunders, J.R. (1984) Nucl. Acids Res. 12: 6701-6708.
216. Old, R., Murray, K. and Roizes, G. (1975) J. Mol. Biol. 92: 331-339.
217. Olson, J.A., Myers, P.A. and Roberts, R.J. unpublished observations.
218. Orekhov, A.V., Rebentish, B.A. and Debavov, V.G. (1982) Dokl. Akad. Nauk. SSSR 263: 217-220.
219. Parker, P. and Schildkraut, I. unpublished observations.
220. Piekarowicz, A. (1982) J. Mol. Biol. 157: 373-381.
221. Piekarowicz, A. unpublished observations.
222. Piekarowicz, A., Bickle, T.A., Shepherd, J.C.W. and Ineichen, K. (1981) J. Mol. Biol. 146: 167-172.
223. Piekarowicz, A., Stasiak, A. and Stanczak, J. (1980) Acta Microbiol. Pol. 29: 151-156.
224. Pirrotta, V. (1976) Nucl. Acids Res. 3: 1747-1760.
225. Pope, A., Lynn, S.P. and Gardner, J.F. unpublished observations.
226. Pugatsch, T. and Weber, H. (1979) Nucl. Acids Res. 7: 1429-1444.
227. Purvis, I.J. and Moseley, B.E.B. (1983) Nucl. Acids Res. 11: 5467-5474.
228. Qiang, B-Q. and Schildkraut, I. unpublished observations.
229. Qiang, B-Q. and Schildkraut, I. (1984) Nucl. Acids Res. 12: 4507-4515.
230. Qiang, B-Q. and Schildkraut, I. and Visentin, L. unpublished observations.
231. Ravetch, J.V., Horiuchi, K. and Zinder, N.D. (1978) Proc. Natl. Acad. Sci. USA 75: 2266-2270.
232. Reaston, J., Duyvesteyn, M.G.C. and de Waard, A. (1982) Gene 20: 103-110.
233. Reister, J. and Yuan, R. (1977) J. Biol. Chem. 252: 451-456.
234. Roberts, R.J. (1985) Nucl. Acids Res. 13: r165-r200.
235. Roberts, R.J., Breitmeyer, J.B., Tabachnik, N.F. and Myers, P.A. (1975) J. Mol. Biol. 91: 121-123.
236. Roberts, R.J., Myers, P.A., Morrison, A., and Murray, K. (1976) J. Mol. Biol. 102: 157-165.
237. Roberts, R.J., Myers, P.A., Morrison, A., and Murray, K. (1976) J. Mol. Biol. 103: 199-208.
238. Roberts, R.J., Wilson, G.A. and Young, F.E. (1977) Nature 265: 82-84.
239. Roizes, G., Nardeux, P-C. and Monier, R. (1979) FEBS Letters 104: 39-44.
240. Roizes, G., Pages, M., Lecou, C., Patillon, M. and Kovoor, A. (1979) Gene 6: 43-50.
241. Roizes, G., Patillon, M. and Kovoor, A. (1977) FEBS Letters 82: 69-70.

242. Rosenvold, E.C. unpublished observations.
243. Rosenvold, E.C. and Szybalski, W. unpublished observations., cited in Gene 7, 217-270 (1979).
244. Roy, P.H. and Smith, H.O. (1973) J. Mol. Biol. 81: 427-444.
245. Roy, P.H. and Smith, H.O. (1973) J. Mol. Biol. 81: 445-459.
246. Sasaki, J. and Yamada, Y. (1984) Agric. Biol. Chem. 48: 3027-3034.
247. Sato, S., Hutchison, C.A. and Harris, J.I. (1977) Proc. Natl. Acad. Sci. USA 74: 542-546.
248. Sato, S., Nakazawa, K. and Shinomiya, T. (1980) J. Biochem. 88: 737-747.
249. Sato, S. and Shinomiya, T. (1978) J. Biochem. 84: 1319-1321.
250. Schildkraut, I. unpublished observations.
251. Schildkraut, I., Banner, D.B., Rhodes, C.S. and Parekh, S. (1984) Gene 27: 327-329.
252. Schildkraut, I. and Christ, C. unpublished observations.
253. Schildkraut, I. and Comb, D. unpublished observations.
254. Schildkraut, I., Grandoni, C. and Comb, D. unpublished observations.
255. Schildkraut, I. and Greenough, L. unpublished observations.
256. Schildkraut, I. and Wise, R. unpublished observations.
257. Schildkraut, I., Wise, R., Borsetti, R. and Qiang, B-Q. unpublished observations.
258. Schmid, K., Thomm, M., Laminet, A., Laue, F.G., Kessler, C., Stetter, K. and Schmitt, R. (1984) Nuc. Acids Res. 12: 2619-2628.
259. Schwabe, G., Posseckert, G. and Klingmuller, W. (1985) Gene 39: 113-116.
260. Sciaky, D. and Roberts, R.J. unpublished observations.
261. Seurinck, J., van de Voorde, A. and van Montagu, M. (1983) Nucl. Acids Res. 11: 4409-4415.
262. Seurinck, J. and van Montagu, M. unpublished observations.
263. Sharp, P.A., Sugden, B. and Sambrook, J. (1973) Biochemistry 12: 3055-3063.
264. Shen, S., Li, Q., Yan, P., Zhou, B., Ye, S., Lu, Y. and Wang, D. (1980) Sci. Sin. 23: 1435-1442.
265. Shibata, T. and Ando, T. (1975) Mol. Gen. Genetics 138: 269-379.
266. Shibata, T. and Ando, T. (1976) Biochim. Biophysis. Acta 442: 184-196.
267. Shibata, T., Ikawa, S., Kim, C. and Ando, T. (1976) J. Bacteriol. 128: 473-476.
268. Shimatake, H. and Rosenberg, M. unpublished observations.
269. Shimotsu, H., Takahashi, H. and Saito, H. (1980) Gene 11: 219-225.
270. Shimotsu, H., Takahashi, H. and Saito, H. (1980) Agric. Biol. Chem. 44: 1665-1666.

271. Shinomiya, T. unpublished observations.
272. Shinomiya, T., Kobayashi, M. and Sato, S. (1980) Nucl. Acids Res. $\underline{8}$: 3275-3285.
273. Shinomiya, T. and Sato, S. (1980) Nucl. Acids Res. $\underline{8}$: 43-56.
274. Sievert, U. and Rosch, A. unpublished observations.
275. Smith, D.I., Blattner, F.R. and Davies, J. (1976) Nucl. Acids Res. $\underline{3}$: 343-353.
276. Smith, H.O. unpublished observations.
277. Smith, H.O. and Nathans, D. (1973) J. Mol. Biol. $\underline{81}$: 419-423.
278. Smith, H.O. and Wilcox, K.W. (1970) J. Mol. Biol. $\underline{51}$: 379-391.
279. Smith, J. and Comb, D. unpublished observations.
280. Smith, L., Blakesley, R. and Chirikjian, J. unpublished observations.
281. Sokolov, N.N., Fitsner, A.B., Anikeitcheva, N.V., Choroshoutina, Yu.B., Samko, O.T., Kolosha, V.O., Fodor, I. and Votrin, I.I. (1985) Molec. Biol. Rep. $\underline{10}$: 159-161.
282. Sparling, R. and Bhatti, A.R. (1984) Microbios $\underline{41}$: 73-79.
283. Stephens, M.A. (1982) J. Bacteriol. $\underline{149}$: 508-514.
284. Stobberingh, E.E., Schiphof, R. and Sussenbach, J.S. (1977) J. Bacteriol. $\underline{131}$: 645-649.
285. Stotz, A. and Philippson, P. unpublished observations.
286. Streips, U. and Golemboski, B. unpublished observations.
287. Sugisaki, H. (1978) Gene $\underline{3}$: 17-28.
288. Sugisaki, H. and Kanazawa, S. (1981) Gene $\underline{16}$: 73-78.
289. Sugisaki, H., Maekawa, Y., Kanazawa, S. and Takanami, M. (1982) Nucl. Acids Res. $\underline{10}$: 5747-5752.
290. Sugisaki, H., Maekawa, Y., Kanazawa, S. and Takanami, M. (1982) Bull. Inst. Chem. Res. Kyoto Univ. $\underline{60}$: 328-335.
291. Sugisaki, H. and Takanami, M. (1973) Nature New Biology $\underline{246}$: 138-140.
292. Sun, D.K. and Yoo, O.J. unpublished observations.
293. Suri, B., Shepherd, J.C.W. and Bickle, T.A. (1984) EMBO J. $\underline{3}$: 575-579.
294. Sussenbach, J.S., Monfoort, C.H., Schiphof, R. and Stobberingh, E.E. (1976) Nucl. Acids Res. $\underline{3}$: 3193-3202.
295. Sussenbach, J.S., Steenbergh, P.H., Rost, J.A., van Leeuwen, W.J. and van Embden, J.D.A. (1978) Nucl. Acids Res. $\underline{5}$: 1153-1163.
296. Sutcliffe, J.G. and Chruch, G.M. (1978) Nucl. Acids Res. $\underline{5}$: 2313-2319.
297. Syddall, R. and Stachow, C. (1985) Biochim. Biophys. Acta $\underline{825}$: 236-243.
298. Szekeres, M. unpublished observations.
299. Takahashi, H. unpublished observations.
300. Takahashi, H., Shimizu, M., Saito, H., Ikeda, Y. and Sugisaki, H. (1979) Gene $\underline{5}$: 9-18.

301. Takanami, M. (1974) Methods in Mol. Biol. 7: 113-133. 302. Timko, J., Horwitz, A.H., Zelinka, J. and Wilcox, G. (1981) J. Bacteriol. 145: 873-877.
303. Tolstoshev, C.M. and Blakesley, R.W. (1982) Nucl. Acids Res. 10: 1-17.
304. Tomassini, J., Roychoudhury, R., Wu, R. and Roberts, R.J. (1978) Nucl. Acids Res. 5: 4055-4064.
305. Torheim, B. unpublished observations.
306. Trautner, T.A. unpublished observations.
307. Tu, C-P.D., Roychoudhury, R. and Wu, R. (1976) Biochem. Biophys. Res. Comm. 72: 355-362.
308. Uporova, T.M., Nikolskaya, I.I., Rubtosova, E.N. and Debov, S.S. (1981) Vestnik Akad. Medicin. Nauk SSSR 2: 21-26.
309. van den Hondel, C.A.M.J.J., van Leen, R.W., van Arkel, G.A., Duyvesteyn, M. and de Waard, A. (1983) FEMS Microbiology Letters 16: 7-12.
310. van Heuverswyn, H. and Fiers, W. (1980) Gene 9: 195-203.
311. van Montagu, M. unpublished observations.
312. van Montagu, M., Sciaky, D., Myers, P.A. and Roberts, R.J. unpublished observations.
313. van Ormondt, H., Lautenberger, J.A., Linn, S. and de Waard, A. (1973) FEBS Letters 33: 177-180.
314. Vanyushin, B.F. and Dobritsa, A.P. (1975) Biochim. Biophys. Acta 407: 61-72.
315. Venegas, A., Vicuna, R., Alonso, A., Valdes, F. and Yudelevich, A. (1980) FEBS Letters 109: 156-158.
316. Venetianer, P. unpublished observations.
317. Visentin, L.P. unpublished observations.
318. Visentin, L.P., Watson R.J., Martin, S. and Zuker, M. unpublished observations.
319. Walter, F., Hartmann, M. and Roth, M. (1978) Abstracts of 12th FEBS Symposium, Dresden
320. Wang, T.-S. (1981) Ko Hsueh Tung Pao 26: 815-817.
321. Wang, R.Y.-H., Shedlarski, J.G., Farber, M.B. Kuebbing, D. and Ehrlich, M. (1980) Biochim. Biophys. Acta 606: 371-385.
322. Wani, A.A., Stephens, R.E., D'Ambrosio, S.M. and Hart, R.W. (1982) Biochim. Biophys. Acta 697: 178-184.
323. Watson, R., Zuker, M., Martin, S.M. and Visentin, L.P. (1980) FEBS Letters 118: 47-50.
324. Watson, R.J., Schildkraut, I., Qiang, B.-Q., Martin, S.M. and Visentin, L.P. (1982) FEBS Letters 150: 114-116.
325. Whang, Y. and Yoo, O.J. unpublished observations.

326. Whitehead, P.R. and Brown, N.L. (1982) FEBS Letters 143: 296-300.
327. Whitehead, P.R. and Brown, N.L. (1983) FEBS Letters 155: 97-102.
328. Whitehead, P.R. and Brown, N.L. (1985) J. Gen. Microbiol. 131: 951-958.
329. Wilson, G.A. and Young, F.E. (1975) J. Mol. Biol. 97: 123-125.
330. Wilson, G.A. and Young, F.E., in D. Schlessinger (ed.), Restriction and modification in the Bacillus subtilis genospecies. (1976) Microbiology 1976). Amer. Soc. Microbiol., Washington, 350-357.
331. Wilson, G.A. and Young, F.E. unpublished observations.
332. Winkler, K. Diploma Dissertation (1979).
333. Winkler, K. and Rosch, A. unpublished observations.
334. Wu, R., King, C.T. and Jay, E. (1978) Gene 4: 329-336.
335. Yamada, Y., Yoshioka, H., Sasaki, J. and Tahara, Y. (1983) J. Gen. Appl. Microbiol. 29: 157-166.
336. Yoon, H., Suh, H., Han, M.H. and Yoo, O.J. (1985) Korean Biochem. J. 18: 82-87.
337. Yoon, H., Suh, H., Kim, K., Han, M.H. and Yoo, O.J. (1985) Korean Biochem. J. 18: 88-93.
338. Yoshida, Y. and Mise, K. unpublished observations.
339. Yoshioka, H., Nakamura, H., Sasaki, J., Tahara, Y. and Yamada, Y. (1983) Agric. Biol. Chem. 47: 2871-2879.
340. Yoshimori, R.N. Ph.D. Thesis (1971).
341. Zabeau, M., Greene, R., Myers, P.A. and Roberts, R.J. unpublished observations.
342. Zabeau, M. and Roberts, R.J. unpublished observations.
343. Zain, B.S. and Roberts, R.J. (1977) J. Mol. Biol. 115: 249-255.
344. Zerler, B., Myers, P.A., Escalante, H. and Roberts, R.J. unpublished observations.
345. Calleja, F., Tandeau de Marsac, N., Coursin, T., van Ormondt, H. and de Waard, A. (1985) Nucl. Acids Res. 13: 6745-6750.
346. Yamada, Y. and Murakami, M. (1985) Agric. Biol. Chem. 49: 3627-3629.
348. Sasaki, J., Murakami, M. and Yamada, Y. (1985) Agric. Biol. Chem. 49: 3107-3122.
350. McClelland, M., Nelson, M. and Cantor, C.R. (1985) Nucl. Acids Res. 13: 7171-7182.
351. Morgan, R. and Hempstead, S.K. unpublished observations.
352. Nagaraja, V., Shepherd, J.C.W., Pripfl, T. and Bickle, T.A. (1985) J. Mol. Biol. 182: 579-587.
353. Nagaraja, V., Shepherd, J.C.W. and Bickle, T.A. (1985) Nature 316: 371-372.
355. Morgan, R., Camp, R. and Soltis, A. unpublished observations.

356. Hiraoka, N., Kita, K., Nakajima, F., Kimizuka, F. and Obayashi, A. (1985) J. Ferment. Technol. 63: 151-157.
357. Kita, K., Hiraoka, N., Oshima, A., Kadonishi, S. and Obayashi, A. (1985) Nucl. Acids Res. 13: 8685-8694.
359. Stote, R. and Schildkraut, I. unpublished observations.
360. Dingman, C. and Schildkraut, I. unpublished observations.
361. Wickberg, L. and Schildkraut, I. unpublished observations.
362. Christ, C. and Wickberg, L. unpublished observations.
363. Walker, J.N.B., Dean, P.D.G. and Saunders, J.R. (1986) Nucl. Acids Res. 14: 1293-1301.
364. Evans, L.R. and Brown, N.L. unpublished observations.
365. Mullings, R., Evans, L.R. and Brown, N.L. unpublished observations.
366. Eastlake, P., unpublished observations.
367. Walker, J.M., Vivian, A. and Saunders, J.R. unpublished observations.
368. Walker, J.M., Dean, P.G. and Saunders, J.R. unpublished observations.
369. Walker, J.M. and Dean, P.G. (1985) Biochem. Soc. Transactions 13: 1055-1058.
370. Kawamura, M., Sakakibara, M., Watanabe, T., Kita, K., Hiraoka, N., Obayashi, A., Takagi, M. and Yano, K. (1986) Nucl. Acids Res. 14: 1985-1989.
371. Qiang, B.-Q. and Schildkraut, I. (1986) Nucl. Acids Res. 14: 1991-1999.
372. Bakh, N.L., Tsvetkova, N.V., Semina, I.E., Tarasov, A.P., Mileikovskaya, M.M., Gruber, I.M., Polyachenko, V.M. and Romanenko, E.E. (1985) Antibiot. Med. Biotek. 30: 342-344.
374. Uporova, T.M., Kartasheva, I.M., Shripkin, E.A., Lopareva, E.N., Nikol'skaya, I.I. and Debov, S.S. (1985) Vopr. Med. Khim. 31: 131-136.
375. Weule, K. and Roberts, R.J. unpublished observations.
376. Kramarov, V.M. and Smolyaninov, V.V. (1981) Biokhimiya 46: 1526-1529.
377. Prangishvili, D.A., Vashakidze, R.P., Chelidze, M.G. and Gabriadze, I.Yu. (1985) FEBS Letters 192: 57-60.
378. Sohail, A., Khan, E., Riazuddin, S. and Roberts, R.J. unpublished observations.
380. Gordon, R. unpublished observations.
381. Bolton, B.J., Comer, M.J. and Kessler, C. unpublished observations.
382. Vasquez, C. (1985) Biochem. Int. 10: 655-662.
383. Walder, R.Y., Walder, J.A. and Donelson, J.E. (1984) J. Biol. Chem. 259: 8015-8026.
384. Butkus, V., Bitinaite, J., Kersulyte, D. and Janulaitis, A. (1985) Biochim. Biophys Acta 826: 208-212.
385. Piekarowicz, A., Goguen, J.D. and Skrzypek, E. (1985) Eur. J. Biochem. 152: 387-393.

387. Price, C. and Bickle, T. unpublished observations.
388. Hall, D., Morgan, R. and Camp, R. unpublished observations.
389. Morgan, R. and Ingalls, D. unpublished observations.
390. Stote, R. and Morgan, R. unpublished observations.
391. Piekarowicz, A. and Goguen, J.D. (1986) Eur. J. Biochem. 154: 295-298.

2

Restriction Endonuclease: Cleavage, Ligation, and Sensitivity

Robert W. Blakesley

Molecular Biology R&D
Bethesda Research Laboratories, Life Technologies, Inc.
8717 Grovemont Circle
Gaithersburg, Maryland 20877

INTRODUCTION

Restriction endonucleases are important tools for the molecular biologist. These enzymes are routinely used to subdivide DNA molecules in a very specific, predictable fashion, allowing one to isolate certain regions for study. DNA sequencing, cloning, mapping, hybridization, and genome characterization are some of the more common procedures incorporating restriction endonuclease treated DNA.

The characteristics of the restriction endonucleases and their reactions have been reviewed [1, 2]. A comprehensive listing of the enzymes and the diversity of microbiological sources appears tin this volume [3]. Informative tables on some characteristics of restriction endonucleases have appeared elsewhere [4, 5]. Tabulated in this chapter are a number of facts frequently used when planning experiments that utilize restriction endonucleases. The final table summarizes a number of general characteristics of the enzyme which should be valuable when designing complex strategies or trouble-shooting unexpected results.

RECOGNITION SEQUENCES

The predictable cleavage of DNA by restriction endonucleases results from their recognition of a certain sequence of nucleotides as substrate. For any individual enzyme this specificity is characteristic, and is essentially invariant between DNAs. Enzymes isolated from different sources usually recognize different nucleotide sequences. The availability of a collection of enzymes with different recognition sequences permits flexibility in dissecting a DNA molecule at any number of specific locations. Even when the sequence of the DNA is unknown, the enzyme's recognition sequence usually implies the end sequence of the cleaved DNA, allowing predictable manipulation, e.g., insertion into a particular site in a cloning vector. For DNAs of known sequence, the recognition sequences can be located and restriction endonuclease cleavage modeled. With such diversity there is also the opportunity to evaluate several alternative manipulation strategies, thereby improving laboratory efficiency and increasing the number of possible outcomes.

The restriction endonucleases have been listed in this volume with their corresponding recognition sequences [3]. In the following table (Table I) the information is arranged rather by recognition sequence. The sequences are grouped by length, from four to twelve nucleotides, and arranged alphabetically within each group. In the case where a particular sequence of DNA is desired, those restriction endonucleases that recognize this sequence are quickly identified from the table. Or it can be found that no known enzyme recognizes the sequence specified. Further, enzymes that, in addition, cleave a methylated form of the sequence, or that cleave a series of closely related (degenerate) sequences are distinguished.

TABLE I. Sequences recognized by restriction endonucleases.

Recognition sequence[a]			Restriction endonuclease[b]	
Unmodified	Methylated	Degenerate	Cleavage site known	Cleavage site unknown
A▼CGT			Mae II	
AG▼CT			Alu I	Oxa I
CATG▼			Nla III	
C▼CGG			Hpa II, Hap II, Mno I	Bsu1192 I, BsuF I, Mni II, SfaGU I
C▼CGG	C▼^{m}CGG		Msp I	
CCTCN₇▼		---[c]	Mnl I	
CG▼CG			FnuD II, Tha I	Acc II, BceF I, BceR I, Bsu1192 II, Bsu1193, Bsu6633, BsuE II, Hin1056 I
C▼TAG			Mae I	Mja I
▼N₇GAGG		---[c]	Mnl I	
▼GATC	▼G^{m}ATC		FnuE I, Sau3A I	BsrP II, Mth I, Pfa I
▼GATC	▼GAT^{m}C		Mbo I	
▼GATC			Bce243, Cpf I, FnuC I	BsaP I, BssG II, BstE III, BstX II, Cpa I, Dpn II, FnuA II, Mno III, Mos I, Nde II, Nfl I, NflA II, Nla II, NsiA I, Nsu I, SinM I
---[d]	G^{m}A▼TC		Dpn I	Cfu I, NmuD I, NmuE I, NsuD I
G▼CGC			HinP I, SciN I	HinS₁ I, HinS₂ I
GCG▼C			FnuD III, Hha I	Cfo I, HinGU I, Mnn IV

TABLE I. (Continued)

Recognition sequence[a]			Restriction endonuclease[b]	
Unmodified	Methylated	Degenerate	Cleavage site known	Cleavage site unknown
GG▼CC	GG▼C^{m}C		Hae III	Blu II, Bse I, BssC I, BstC I,
GG▼CC			BspR I, BsuR I,	Bsu1076, Bsu114, Clm I,
			Clt I, FnuD I,	Hhg I, Mni I, Mnn II, Ngo II,
			Sfa I	Nla I, Pai I, Pal I, Ppu I,
				Sul I, Ttn I, Vha I
GT▼AC			Rsa I	
T▼CGA	T▼^{m}CGA		Taq I	
TCGA				Tfl I, TthHB8 I
▼N$_{13}$CATCC	---[c]		Fok I	HinGU II
CCAGG	▼^{m}CCAGG	CCWGG	EcoR II	
CC▼AGG	^{m}CC▼AGG	CCWGG	BstN I, TaqX I	
CC▼AGG	C^{m}C▼AGG	CCWGG	Apy I, BstN I	
CC▼AGG	^{m}CC▼AGG	CCNGG	ScrF I	
CC▼AGG		CCWGG	Aor I, Mva I	Atu II, AtuB I, BinS I, BstG II,
				Cfr5 I, CfrlI I, Eca II, Ecl II,
				Eco27 I, Eco38 I, Mph I, SgrI I
CC▼CGG	^{m}CC▼CGG	CCNGG	ScrF I	
CC▼CGG	^{m}CCC▼GG	CCSGG	Cau II, Nci I	Aha I, Rsh II
CCC▼GG		CCSGG	Bcn I	
CC▼GGG	^{m}CC▼GGG	CCNGG	ScrF I	
CC▼GGG	^{m}CCG▼GG	CCSGG	Cau II, Nci I	Aha I, Rsh II
CCG▼GG		CCSGG	Bcn I	

55

TABLE I. (Continued)

Recognition sequence[a)]			Restriction endonuclease[b)]	
Unmodified	Methylated	Degenerate	Cleavage site known	Cleavage site unknown
▼CCTGG	▼mCCTGG	CCWGG	EcoR II	
CC▼TGG	mCC▼TGG	CCWGG	BstN I, TaqX I	
CC▼TGG	Cm̄C▼TGG	CCWGG	Apy I, BstN I	
CC▼TGG	Cm̄C▼TGG	CCNGG	ScrF I	
CC▼TGG	m̄CC▼TGG	CCWGG	Aor I, Mva I	Atu II, AtuB I, BinS I, BstG II, Cfr5 I, Cfr11 I, Eca II, Ecl II, Eco27 I, Eco38 I, Mph I, SgrI I
C▼TAAG		CTNAG	Dde I	
C▼TCAG		CTNAG	Dde I	
C▼TGAG		CTNAG	Dde I	
C▼TTAG		CTNAG	Dde I	
GAAGAN₈▼		---[c)]	Mbo II	Ncu I, Tce I
G▼AATC		GANTC	FnuA I, Hinf I	Hha II, Nca I, Nov II, NsiH I
GACGCN₅▼		---[c)]	Hga I	
G▼ACTC		GANTC	FnuA I, Hinf I	Hha II, Nca I, Nov II, NsiH I
G▼AGTC		GANTC	FnuA I, Hinf I	Hha II, Nca I, Nov II, NsiH I
▼N₅GATCC		---[c)]	Bin I	Bth II
▼N₉GATGC	▼N₉GATG^{m}C	---[c)]	SfaN I	
G▼ATTC		GANTC	FnuA I, Hinf I	Hha II, Nca I, Nov II, NsiH I

TABLE I. (Continued)

Recognition sequence[a]			Restriction endonuclease[b]	
Unmodified	Methylated	Degenerate	Cleavage site known	Cleavage site unknown
GC▼AGC		GCNGC	Fnu4H I	
GCAGCN$_8$▼		GCWGC	Bbv I	
GCATCN$_5$▼		---[c]	SfaN I	
GC▼CGC		GCNGC	Fnu4H I	
GC▼GGC		GCNGC	Fnu4H I	
▼N$_{10}$GCGTC		---[c]	Hga I	
GC▼TGC		GCNGC	Fnu4H I	
▼N$_{12}$GCTGC		GCWGC	Bbv I	
G▼GACC		GGWCC	Afl I, Ava II, BamN$_x$ I, Cau I, FdiV I, HgiB I, HgiC II, HgiE I, HgiH III	Asp697, Bme216, Bti I, Clm II, Eco47 I, NspH II, Sin I
G▼GACC		GGNCC	Asu I, Nsp7524 IV, Sau96 I	Cfr4 I, Cfr8 I, Cfr13 I, Eco39 I, Eco47 II, Mja II, NmuS I, NmuE II, Psp I, Sdy I
GGATCN$_4$▼		---[c]	Bin I	Bth II
GGATGN$_9$▼		---[c]	Fok I	HinGU II
G▼GCCC		GGNCC	Asu I, Nsp7524 IV, Sau96 I	Cfr4 I, Cfr8 I, Cfr13 I, Eco39 I, Eco47 II, Mja II, NmuS I, NmuE II, Psp I, Sdy I

TABLE I. (Continued)

Recognition sequence[a]			Restriction endonuclease[b]	
Unmodified	Methylated	Degenerate	Cleavage site known	Cleavage site unknown
G▼GGCC		GGNCC	Asu I, Nsp7524 IV, Sau96 I	Cfr4 I, Cfr8 I, Cfr13 I, Eco39 I, Eco47 II, Mja II, NmuS I, NmuE II, Psp I, Sdy I
G▼GTCC		GGWCC	Afl I, Ava II, Bam Nx, Cau I, Fdi I, HgiB I, HgiC II, HgiE I, HgiH III	Asp697, Bme216, Bti I, Clm I, Eco47I, NspH II, Sin I
G▼GTCC		GGNCC	Asu I, Nsp7524 IV, Sau96 I	Cfr4 I, Cfr8 I, Cfr13 I, Eco39 I, Eco47 II, Mja II, NmuS I, NmuE II, Psp I, Sdy I
GGTGAN$_8$▼		---[c]	Hph I	
▼GTAAC		GTNAC	Mae III	
▼GTCAC		GTNAC	Mae III	
▼GTGAC		GTNAC	Mae III	
▼GTTAC		GTNAC	Mae III	
▼N$_7$TCACC		---[c]	Hph I	
▼N$_7$TCTTC		---[c]	Mbo II	Ncu I, Tce I
A▼AGCTT			Eco VIII, Hind III, Hsu I	Bbr I, Bpe I, Chu I, Hin 173, Hinb III, Hinf II, HinJC II, Mki I
AAT▼ATT			Ssp I	

TABLE I. (Continued)

Recognition sequence[a]			Restriction endonuclease[b]	
Unmodified	Methylated	Degenerate	Cleavage site known	Cleavage site unknown
A▼CACGT		ACRYGT	Afl III	
ACATG▼C		RCATGY	Nsp7524 I, NspH I	
A▼CATGT ACATG▼T		ACRYGT RCATGY	Afl III Nsp7524 I, NspH I	
ACCGGC		RCCGGY		Cfr10 I
ACCGGT		RCCGGY		Cfr10 I
ACCTGCN₄▼		---[c]	BspM I	
A▼CGCGT A▼CGCGT	mA▼CGCGT	ACRYGT	Afl III Mlu I	
A▼CGTGT		ACRYGT	Afl III	
A▼CTAGT			Spe I	
A▼GATCC	A▼GmATCC	RGATCY	Xho II	
A▼GATCT A▼GATCT	A▼GmATCT A▼GmATCT	RGATCY	Xho II Bgl II	
AGCGC▼C		RGCGCY	Hae II	
AGCGC▼T AGCGCT		RGCGCY	Hae II	HinH I, Ngo I HinH I, Ngo I Eco47 III
AGG▼CCT			Gdi I, Stu I	Aat I

TABLE I. (Continued)

Recognition sequence[a]			Restriction endonuclease[b]	
Unmodified	Methylated	Degenerate	Cleavage site known	Cleavage site unknown
AGT▼ACT			Sca I	
AT▼CGAT			Cla I	Asp707, Ban III
ATGCA▼T			Nsi I	Ava III
CAAACAN₁₁▼		---[c]	Tth111 II	
CAAGCAN₁₁▼		---[c]	Tth111 II	
CACCCAN₁₁▼		---[c]	Taq II	
CAG▼CGG		CMGCKG	NspB II	
CAG▼CTG		CMGCKG	NspB II	
CAG▼CTG			Pvu II	Cfr6 I
CA▼TATG			Nde I	
C▼CAAGG		CCNNGG	Sec I	
C▼CAAGG		CCWWGG	Sty I	EcoT14, EcoT104
C▼CACGG		CCNNGG	Sec I	
C▼CAGGG		CCNNGG	Sec I	
C▼CATGG		CCNNGG	Sec I	
C▼CATGG		CCWWGG	Sty I	EcoT14, EcoT104
C▼CATGG			Nco I	
C▼CCAGG		CCNNGG	Sec I	

TABLE I. (Continued)

Recognition sequence[a]			Restriction endonuclease[b]	
Unmodified	Methylated	Degenerate	Cleavage site known	Cleavage site unknown
C▼CCCGG		CCNNGG	Sec I	
C▼CCGAG		CYCGRG	Ava I, Nsp7524 III	Aqu I, Avr I
C▼CCGGG	C▼mCCGGG	CCNNGG	Sec I	
C▼CCGGG		CYCGRG	Ava I	
C▼CCGGG		CYCGRG	Nsp7524 III	Aqu I, Avr I
C▼CCGGG		CYCGRG	Cfr9 I, Xcy I, Xma I	
CCC▼GGG			Sma I	
C▼CCTGG		CCNNGG	Sec I	
C▼CGAGG		CCNNGG	Sec I	
C▼CGCGG		CCNNGG	Sec I	
CCG▼CGG		CMGCKG	NspB II	
CCGC▼GG			Csc I, Sac II, Sst II	Bac I, Ecc I, Mra I, Ngo III, NlaS I, Sbo I, Sfr I, Shy I, Tgl I
CCG▼CTG		CMGCKG	NspB II	
C▼CGGGG		CCNNGG	Sec I	
C▼CGTGG		CCNNGG	Sec I	
C▼CTAGG		CCNNGG	Sec I	
C▼CTAGG		CCWWGG	Sty I	EcoT14, EcoT104
C▼CTAGG			Avr II	
C▼CTCGG		CCNNGG	Sec I	

TABLE I. (Continued)

Recognition sequence[a]			Restriction endonuclease[b]	
Unmodified	Methylated	Degenerate	Cleavage site known	Cleavage site unknown
C▼CTGGG		CCNNGG	Sec I	
C▼CTTGG		CCNNGG	Sec I	
C▼CTTGG		CCWWGG	Sty I	EcoT14, EcoT104
CGAT▼CG	CG^{m}AT▼CG		Pvu I, Xor II	
CGAT▼CG			Nbl I, Rsh I	Rsp I, Xni I
C▼GGCCA		YGGCCR	Cfr I, Cfr 14 I, Eae I	
C▼GGCCG		YGGCCR	Cfr I, Cfr14 I, Eae I	
C▼GGCCG		YGGCCG	Gdi II	
C▼GGCCG			Eag I, Xma III	
CTCCAG		---[c]		Gsb I, Gsu I
C▼TCGAG		CYCGRG	Ava I, Nsp7524 III	Aqu I, Avr I
C▼TCGAG			Blu I, PaeR7, Pan I, Sla I, Xho I, Xpa I	Abr I, Asp703, Bbi III, BssH I, BstH I, BsuM, Bth I, Ccr II, Dde II, Mca I, Msi I, PflW I, Spa I, Scu I, Sex I, Sga I, Sgo I, Slu I
C▼TCGGG		CYCGRG	Ava I, Nsp7524 III	Aqu I, Avr I
CTGCA▼G			Pst I, SalP I, Sfl I	Asp708, Bbi I, Bce170, Bsu1247, Cau III EaeP I, Eco36 I, Eco48 I, Eco49 I, Mau I, Noc I, Pma I, Ska II, Xma II Xor I

TABLE I. (Continued)

Recognition sequence[a]			Restriction endonuclease[b]	
Unmodified	Methylated	Degenerate	Cleavage site known	Cleavage site unknown
CTGGAG		---[c]		Gsb I, Gsu I
C▼TTAAG			Afl II	
GAATGCN▼		---[c]	Bsm I	
G▼AATTC GAATTC	G▼mAATTC		EcoR I	
GACCGAN₁₁▼		---[c]	Taq II	Rsr I
GA▼CGCC		GRCGYC	Acy I, Aha II, Aos II, AstW I, Asu III, HgiD I, HgiG I, HgiH II	Bbi II, NlaS II
GA▼CGTC		GRCGYC	Acy I, Aha II, Aos II, AstW I, Asu III, HgiD I, HgiG I, HgiH II Aat II	Bbi II, NlaS II
GACGT▼C				
GAGCA▼C		GDGCHC	Bsp1286 I, Nsp7524 II	Sdu I
GAGCA▼C		GWGCWC	HgiA I	
GAGCC▼C		GDGCHC	Bsp1286 I, Nsp7524 II	Sdu I
GAGCC▼C		GRGCYC	Ban II, Bvu I, HgiJ II	Eco24 I, Eco25 I, Eco26 I, Eco35 I, Eco40 I, Eco41 I

TABLE I. (Continued)

Recognition sequence[a]			Restriction endonuclease[b]	
Unmodified	Methylated	Degenerate	Cleavage site known	Cleavage site unknown
GAGCT▼C		GDGCHC	Bsp1286 I, Nsp7524 II	Sdu I
GAGCT▼C		GRGCYC	Ban II, Bvu I, HgiJ II	Eco24 I, Eco25 I, Eco26 I, Eco35 I, Eco40 I, Eco41 I
GAGCT▼C GAGCT▼C GAGCT▼C	G^{m}AGCT▼C	GWCWC	HgiA I Sac I Sst I	EcoICR I
GAT▼ATC			EcoR V	Eco32 I, NflA I
▼N$_8$GCAGGT		---[c]	BspM I	
GCATG▼C GCATG▼C GCATGC	GCATG▼^{m}C	RCATGY	Nsp7524 I, NspH I Sph I	Spa I
GCATG▼T		RCATGY	Nsp7524 I, NspH I	
G▼CATTC		---[c]	Bsm I	
GCC▼GGC			Nae I	Mis I, Nba I, Nbr I, Nmu I, NmuF I, Pgi I, Rlu I, Ska I Cfr10 I
GCCGGC		RCCGGY		Cfr10 I
GCCGGT		RCCGGY		
G▼CGCGC			BssH II	BseP I, BsrH I, BsoP I
G▼CTAGC			Nhe I	
GGA▼ACC		GGNNCC	Nla IV	

TABLE I. (Continued)

Recognition sequence[a]			Restriction endonuclease[b]	
Unmodified	Methylated	Degenerate	Cleavage site known	Cleavage site unknown
GGA▼CCC		GGNNCC	Nla IV	
GGA▼GCC		GGNNCC	Nla IV	
GGA▼TCC		GGNNCC	Nla IV	
G▼GATCC	G▼G^{m}ATCC	RGATCY	Xho II	
G▼GATCC	G▼G^{m}ATCC		BamH I	
G▼GATCC	G▼GATC^{m}C		BamH I	
G▼GATCC			Ali I, Bst I	Aac I, Aae I, BamF I, BamK I, BamN I, Dds I, Gdo I, Gin I, Gox I, Rhs I
G▼GATCT	G▼G^{m}ATCT	RGATCY	Xho II	
GGC▼ACC		GGNNCC	Nla IV	
G▼GCACC		GGYRCC	Ban I, HgiH I	Eco50 I
▼GGCACC		GGYRCC	HgiC I	
GGC▼CCC		GGNNCC	Nla IV	
GGC▼GCC		GGNNCC	Nla IV	
G▼GCGCC		GGYRCC	Ban I, HgiH I	Eco50 I
▶GGCGCC		GGYRCC	HgiC I	
GG▼CGCC		GRCGYC	Acy I, Aha II, Aos II, AstW I, Asu III, HgiD I, HgiG I, HgiH I	Bbi II, NlaS II
GGGCG▼C		RGCGCY	Hae II	HinH I, Ngo I
GG▼CGCC	GG▼CGC^{m}C		Nar I	
GG▼CGCC			Nda I, Nun II	BbeA I, BinS II, Nam I
GGCGC▼C			Bbe I	
GGCGC▼T		RGCGCY	Hae II	HinH I, Ngo I

65

TABLE I. (Continued)

Recognition sequence[a]			Restriction endonuclease[b]	
Unmodified	Methylated	Degenerate	Cleavage site known	Cleavage site unknown
GG▼CGTC		GRCGYC	Acy I, Aha II, Aos II, AstW I, Asu III, HgiD I, HgiG I, HgiH I	Bbi II, NlaS II
GGC▼TCC		GGNNCC	Nla IV	
GGG▼ACC		GGNNCC	Nla IV	
GGGCA▼C		GDGCHC	Bsp1286 I, Nsp7524 II	Sdu I
GGG▼CCC		GGNNCC	Nla IV	
GGGCC▼C		GDGCHC	Bsp1286 I, Nsp7524 II	Sdu I
GGGCC▼C		GRGCYC	Ban II, Bvu I, HgiJ II, Apa I	Eco24 I, Eco25 I, Eco26 I, Eco35 I, Eco40 I, Eco41 I
GGGCC▼C				
GGGCT▼C		GDGCHC	Bsp1286 I, Nsp7524 II	Sdu I
GGGCT▼C		GRGCYC	Ban II, Bvu I, HgiJ II	Eco24 I, Eco25 I, Eco26 I, Eco35 I, Eco40 I, Eco41 I
GGG▼GCC		GGNNCC	Nla IV	
GGG▼TCC		GGNNCC	Nla IV	

TABLE I. (Continued)

Recognition sequence[a]			Restriction endonuclease[b]	
Unmodified	Methylated	Degenerate	Cleavage site known	Cleavage site unknown
GGT▼ACC		GGNNCC	Nla IV	
G▼GTACC		GGYRCC	Ban I, HgiH I	Eco50 I
▼GGTACC		GGYRCC	HgiCI	
GGTAC▼C	GGTA^{m}C▼C		Kpn I	
GGTAC▼C	GGTAC▼^{m}C		Kpn I	
G▼GTACC			Asp718	Nmi I
GGT▼CCC		GGNNCC	Nla IV	
GGT▼GCC		GGNNCC	Nla IV	
G▼GTGCC		GGYRCC	Ban I, HgiH I	Eco50 I
▼GGTGCC		GGYRCC	HgiC I	
GGT▼TCC		GGNNCC	Nla IV	
GT▼AGAC		GTMKAC	Acc I	
GT▼ATAC		GTMKAC	Acc I	
GTATAC				Sna I
GTC▼AAC		GTYRAC	Hind II, HinJC I	Chu II, Hinc II, Mnn I
GT▼CGAC		GTMKAC	Acc I	
GTC▼GAC		GTYRAC	Hind II, HinJC I	Chu II, Hinc II, Mnn I
G▼TCGAC			HgiC III, HgiD II, Nop I, Sal I	Rhe I, Rhp I, Rrh I, Rro I, Xam I
GT▼CTAC		GTMKAC	Acc I	
GTGCA▼C		GDGCHC	Bsp1286 I, Nsp7524 II	Sdu I
GTGCA▼C		GWGCWC	HgiA I	

TABLE I. (Continued)

Recognition sequence[a]			Restriction endonuclease[b]	
Unmodified	Methylated	Degenerate	Cleavage site known	Cleavage site unknown
GTGCC▼C		GDGCHC	Bsp1286 I, Nsp7524 II	Sdu I
GTGCT▼C		GDGCHC	Bsp1286 I, Nsp7524 II	Sdu I
GTGCT▼C		GWGCWC	HgiA I	
GTT▼AAC GTT▼AAC GTTAAC	GTT▼AA^{m}C	GTYRAC	Hind II, HinJC I Hpa I	Chu II, Hinc II, Mnn I Bse II
GTT▼GAC		GTYRAC	Hind II, HinJC I	Chu II, Hinc II, Mnn I
TAC▼GTA			SnaB I	
T▼CCGGA			Acc III, BspM II	
TCG▼CGA			Nru I	Ama I, Sbo13
▼N$_9$TCGGTC		---[c]	Taq II	
T▼CTAGA			Xba I	
T▼GATCA TGATCA	T▼GAT^{m}CA		Bcl I	AtuC I, BstG I, Cpe I, Sst IV
TGC▼GCA			Aos I, Fdi II, Mst I	Fsp I
▼N$_9$TGCTTG		---[c]	Tth111 II	

TABLE I. (Continued)

Recognition sequence[a]			Restriction endonuclease[b]	
Unmodified	Methylated	Degenerate	Cleavage site known	Cleavage site unknown
T▼GGCCA		YGGCCR	Cfr I, Cfr14 I, Eae I	
TGG▼CCA			Bal I	
T▼GGCCG		YGGCCG	Gdi II	
T▼GGCCG		YGGCCR	Cfr I, Cfr14 I, Eae I	
▼N$_9$TGGTG		---[c]	Taq II	
▼N$_9$TGTTTG		---[c]	Tth111 II	
TT▼CGAA			Asu II, Fsp II, Mla I	Nsp7524 V, NspB I
TTT▼AAA			Aha III, Dra I	
AG▼GACCC		RGGNCCY	Dra II, EcoO109 I	
AG▼GACCC		RGGWCCY	PpuM I	
AG▼GACCT		RGGNCCY	Dra II, EcoO109 I	
AG▼GACCT		RGGWCCY	PpuM I	
AG▼GCCCC		RGGNCCY	Dra II, EcoO109 I	
AG▼GCCCT		RGGNCCY	Dra II, EcoO109 I	
AG▼GGCCC		RGGNCCY	Dra II, EcoO109 I	
AG▼GGCCT		RGGNCCY	Dra II, EcoO109 I	

TABLE I. (Continued)

Recognition sequence[a]			Restriction endonuclease[b]	
Unmodified	Methylated	Degenerate	Cleavage site known	Cleavage site unknown
AG▼GTCCC AG▼GTCCC		RGGNCCY RGGWCCY	Dra II, EcoO109 I PpuM I	
AG▼GTCCT AG▼GTCCT		RGGNCCY RGGWCCY	Dra II, EcoO109 I PpuM I	
CC▼TAAGG CC▼TAAGG	mCC▼TAAGG	CCTNAGG CCTNAGG	Mst II Aoc I, Axy I, Cvn I, Sau I	
CC▼TCAGG	mCC▼TCAGG	CCTNAGG	Mst II	
CC▼TCAGG		CCTNAGG	Aoc I, Axy I, Cvn I, Sau I	
CC▼TGAGG CC▼TGAGG	mCC▼TGAGG	CCTNAGG CCTNAGG	Mst II Aoc I, Axy I, Cvn I, Sau I	
CC▼TTAGG CC▼TTAGG	mCC▼TTAGG	CCTNAGG CCTNAGG	Mst II Aoc I, Axy I, Cvn I, Sau I	
CG▼GACCG		CGGWCCG	Rsr II	
CG▼GTCCG		CGGWCCG	Rsr II	
GC▼TAAGC		GCTNAGC	Esp I	
GC▼TCAGC		GCTNAGC	Esp I	
GC▼TGAGC		GCTNAGC	Esp I	

TABLE I. (Continued)

Recognition sequence[a]			Restriction endonuclease[b]	
Unmodified	Methylated	Degenerate	Cleavage site known	Cleavage site unknown
GC▼TTAGC		GCTNAGC	Esp I	
GG▼GACCC		RGGNCCY	Dra II, EcoO109 I	
GG▼GACCC		RGGWCCY	PpuM I	
GG▼GACCT		RGGNCCY	Dra II, EcoO109 I	
GG▼GACCT		RGGWCCY	PpuM I	
GG▼GCCCC		RGGNCCY	Dra II, EcoO109 I	
GG▼GCCCT		RGGNCCY	Dra II, EcoO109 I	
GG▼GGCCC		RGGNCCY	Dra II, EcoO109 I	
GG▼GGCCT		RGGNCCY	Dra II, EcoO109 I	
GG▼GTCCC		RGGNCCY	Dra II, EcoO109 I	
GG▼GTCCC		RGGWCCY	PpuM I	
GG▼GTCCT		RGGNCCY	Dra II, EcoO109 I	
GG▼GTCCT		RGGWCCY	PpuM I	
G▼GTAACC	G▼GTAA^{m}C^{m}C	GGTNACC	BstE II	
G▼GTAACC		GGTNACC	AspA I, BstP I, Eca I	Cfr7 I
G▼GTCACC	G▼GTCA^{m}C^{m}C	GGTNACC	BstE II	
G▼GTCACC		GGTNACC	AspA I, BstP I, Eca I	Cfr7 I

TABLE I. (Continued)

Recognition sequence[a]			Restriction endonuclease[b]	
Unmodified	Methylated	Degenerate	Cleavage site known	Cleavage site unknown
G▼GTGACC G▼GTGACC	G▼GTGA^{m}C^{m}C	GGTNACC GGTNACC	BstE II AspA I, BstP I, Eca I	Cfr7 I
G▼GTTACC G▼GTTACC	G▼GTTA^{m}C^{m}C	GGTNACC GGTNACC	BstE II AspA I, BstP I, Eca I	Cfr7 I
GC▼GGCCGC	GC▼GGCCG^{m}C		Not I	
CACN$_3$▼GTG		CACN$_3$GTG	Dra III	
GACN▼N$_2$GTC		GACN$_3$GTC	Tth111 I	Tte I, Ttr I
GAAN$_2$▼N$_2$TTC GAAN$_2$▼N$_2$TTC	GA^{m}AN$_2$▼N$_2$TTC	GAAN$_4$TTC GAAN$_4$TTC	Xmn I Asp700	
GCCN$_4$▼NGGC	GC^{m}CN$_4$▼NGGC	GCCN$_5$GGC	Bgl I	
ACCN$_6$GGT		ACCN$_6$GGT		HgiE II
CCAN$_5$▼NTGG		CCAN$_6$TGG	BstX I	BssG I, BstT I
GGCCN$_4$▼NGGCC GGCCN$_4$▼NGGCC	GG^{m}CCN$_4$▼NGG^{m}CC GGCCN$_4$▼NGGC^{m}C	GGCCN$_5$GGCC GGCCN$_5$GGCC	Sfi I Sfi I	

a) The restriction endonuclease recognition sequences are grouped by increasing length, i.e., tetranucleotides, pentanucleotides, etc. Within each size group, the sequences are listed alphabetically and written 5'→3'. The complementary strand sequence is omitted for brevity. A corresponding methylated sequence is listed in the second column only if the same enzyme also cleaves the modified sequence. Some enzymes recognize a number of related 'palindromic' sequences, e.g., containing purine/pyrimidine 'wobbles'. The generalized or degenerate sequence that summarizes this group of sequences is given in column three.

The information for compiling this table was taken from Roberts [6], McClelland and Nelson [7], Celleja et al. [8], Butkus et al. [9], Mise and Nakajima [10, 11], Kita et al. [12], and New England BioLabs Catalog Update (1/86).

The symbols used in the recognition sequences are the following: A, adenine; C, cytosine; G, guanine; T, thymine; A, N^6-methyladenine; ^{m}C, 5-methylcytosine; and ▼, the location of cleavage. Other code letters representing multiple bases at a given position follow the recommendation of Cornish-Bowden [13]: B = C, G, or T; D = A, G, or T; H = A, C or T; K = G or T; M = A or C; N = A, C, G, or T; R = A or G; S = C or G; V = A, C, or G; W = A or T; and Y = C or T.

b) Restriction endonucleases are listed in two groups for each recognition sequence. Those that cleave the DNA at a known location as indicated by the arrow are placed in the fourth column. Those enzymes known to recognize a particular sequence, but their exact cleavage location was not determined are found in the last column.

c) This recognition sequence is not palindromic, but it is a member of related sequences. In many cases, the related sequence is the reverse complement of that listed.

d) These enzymes recognize and cleave only the methylated sequence given in column two.

CLEAVAGE SEQUENCE PAIRS COMPATABLE FOR LIGATION

The cleavage of a DNA by a particular restriction endonuclease generates DNA ends of defined sequence. Knowing the end sequence, one can determine whether that end will be compatable, i.e., will readily ligate, with an end of a second DNA. This is important, for example, in planning the cloning of DNAs into certain locations of vectors. Table II lists alphabetically the common restriction endonucleases each together with only those enzymes which generate end sequences that are compatable. In some cases a number of possible (degenerate) sequences are possible, but only the fraction indicated are actually compatable. Once cloned, DNA inserts may be excised by cleavage with the enzymes listed in the fourth column. By comparison of several possible cleavage and ligation combinations, the best approach for meeting particular experimental needs can be determined.

TABLE II. Cleavage sequence pairs compatible for ligation.

Ligation of[a]			
Donor end	Acceptor end	Recombined sequence	Recleaved by
Aat II	Aat II	GACGT/C	Aat II, Aha II, Mae II
Acc I[b]	Acc I[b]	GT/MKAC	Acc I, Hind II[g], Sal I[g], Taq I[g]
	Aha II	GT/CGYC	---
	Asu II	GT/CGAA	Taq I
	Cla I	GT/CGAT	Taq I
	HinP I	GT/CGC	---
	Hpa II	GT/CGG	---
	Mae II	GT/CGT	---
	Nar I	GT/CGCC	---
	Taq I	GT/CGA	Taq I
Afl II	Afl II	C/TTAAG	Afl II
Afl III[b]	Afl III[b]	A/CRYGT	Afl III, Mae II[f], Mlu I[g], Nla III[g], Nsp I[g], Tha I[g]
	BssH II	A/CGCGC	Hha I, HinP I, Tha I
	Mlu I	A/CGCGT	Afl III, Mlu I, Tha I
	Nco I	A/CATGG	Nla III
	Sec I[e]	A/CRYGG	Mae II[g], Nla III[g], Tha I[g]
	Sty I[b]	A/CATGG	Nla III
Aha II	Acc I[b]	GR/CGAC	---
	Aha II	GR/CGYC	Aat II[g], Aha II, Ban I[g], Hae II[g], Hga I[f], Hha I[g], HinP I[g], Mae II[g], Nar I[g], Nla IV[g]
	Asu II	GR/CGAA	---
	Cla I	GR/CGAT	---
	HinP I	GR/CGC	Hga I[f], Hha I[f], HinP I[f]
	Hpa II	GR/CGG	---
	Mae II	GR/CGT	Mae II[f]
	Nar I	GR/CGCC	Aha II, Ban I[f], Hae II[f], Hga I[f], Hha I[f], HinP I[f], Nar I[f], Nla IV[f]
	Taq I	GR/CGA	---

TABLE II. (Continued)

Ligation of[a]			
Donor end	Acceptor end	Recombined sequence	Recleaved by
Alu I	Alu I	AG/CT	Alu I
	EcoR V	AG/ATC	Sau3A I
	Nla IV[p]	AG/GCC	Hae III
	NspB II[q]	AG/CTG	Alu I
	Pvu II	AG/CTG	Alu I
Apa I	Apa I	GGGCC/C	Apa I, Ban II, Bsp1286, Hae III, Nla IV, Sau96 I
	Ban II[b]	GGGCC/C	Apa I, Ban II, Bsp1286, Hae III, Nla IV, Sau96 I
	Bsp1286[d]	GGGCC/C	Apa I, Ban II, Bsp1286, Hae III, Nla IV, Sau96 I
Asp718	Asp718	G/GTACC	Asp718, Ban I, Kpn I, Nla IV, Rsa I
	Ban I[b]	G/GTACC	Asp718, Ban I, Kpn I, Nla IV, Rsa I
Asu II	Acc I[b]	TT/CGAC	Taq I
	Aha II	TT/CGYC	---
	Asu II	TT/CGAA	Asu II, Taq I
	Cla I	TT/CGAT	Taq I
	HinP I	TT/CGC	---
	Hpa II	TT/CGG	---
	Mae II	TT/CGT	---
	Nar I	TT/CGCC	---
	Taq I	TT/CGA	Taq I
Ava I[b]	Ava I[b]	C/YCGRG	Ava I, Bcn I[g], Cau II[g], Hpa II[g], ScrF I[g], Sec I[g], Sma I[g], Taq I[g], Xho I[g], Xma I[g]
	BspM II	C/CCGGA	Bcn I, Cau II, Hpa II, ScrF I
	Sal I	C/TCGAC	Taq I
	Sec I[e]	C/CCGGG	Ava I, Bcn I, Cau II, Hpa II, ScrF I, Sec I, Sma I, Xma I
	Xho I	C/TCGAG	Ava I, Taq I, Xho I
	Xma I	C/CCGGG	Ava I, Bcn I, Cau II, Hpa II, ScrF I, Sec I, Sma I, Xma I
Ava II[c]	Ava II[c]	G/GWCC	Ava II, Sau96 I
	Dra II[b]	G/GWCCY	Ava II, Nla IV[f], Sau96 I
	PpuM I[c]	G/GWCCY	Ava II, Nla IV[f], Sau96 I
	Rsr II[c]	G/GWCCG	Ava II, Sau96 I
	Sau96 I[b]	G/GWCC	Ava II, Sau96 I
Avr II	Avr II	C/CTAGG	Avr II, Mae I, Sec I, Sty I
	Nhe I	C/CTAGC	Mae I
	Sec I[e]	C/CTAGG	Avr II, Mae I, Sec I, Sty I
	Spe I	C/CTAGT	Mae I
	Sty I[b]	C/CTAGG	Avr II, Mae I, Sec I, Sty I
	Xba I	C/CTAGA	Mae I

TABLE II. (Continued)

Donor end	Ligation of[a] Acceptor end	Recombined sequence	Recleaved by
Bal I	Bal I	TGG/CCA	Bal I, Cfr I, Hae III
	EcoR V	TGG/ATC	Bin I, Sau3A I
	Hae III	TGG/CC	Hae III
	Nla IV	TGG/NCC	Ava II[f], Hae III[f], Sau96 I
	Stu I	TGG/CCT	Hae III
BamH I	BamH I	G/GATCC	BamH I, Bin I, Nla IV, Sau3A I, Xho II
	Bcl I	G/GATCA	Bin I, Sau3A I
	Bgl II	G/GATCT	Bin I, Sau3A I, Xho II
	Sau3A I	G/GATC	Bin I, Sau3A I
	Xho II	G/GATCY	BamH I[f], Bin I, Nla IV[f], Sau3A I, Xho II
Ban I[b]	Asp718	G/GTACC	Asp718, Ban I, Kpn I, Nla IV, Rsa I
	Ban I[b]	G/GYRCC	Aha II[g], Asp718[g], Ban I, Hae II[g], Hha I[g], HinP I[g], Kpn I[g], Nar I[g], Nla IV, Rsa I[g]
Ban II[b]	Apa I	GGGCC/C	Apa I, Ban II, Bsp1286, Hae III, Nla IV, Sau96 I,
	Ban II[b]	GRGCY/C	Alu I[g], Apa I[g], Ban II, Bsp1286, Hae III[g], HgiA I[g], Nla IV[g], Sau96 I[g], Sst I[g]
	Bsp1286[d]	GRGCY/C	Alu I[g], Apa I[g], Ban II, Bsp1286, Hae III[g], HgiA I[g], Nla IV[g], Sau96 I[g], Sst I[g]
	HgiA I[b]	GAGCT/C	Alu I, Ban II, Bsp1286, HgiA I, Sst I
	Sst I	GAGCT/C	Alu I, Ban II, Bsp1286, HgiA I, Sst I
Bcl I	BamH I	T/GATCC	Bin I, Sau3A I
	Bcl I	T/GATCA	Bcl I, Sau3A I
	Bgl II	T/GATCT	Sau3A I
	Sau3A I	T/GATC	Sau3A I
	Xho II	T/GATCY	Bin I[f], Sau3A I
Bcn I[c]	Bcn I[c]	CCS/GG	Bcn I, Cau II, Hpa II, ScrF I
Bgl II	BamH I	A/GATCC	Bin I, Sau3A I, Xho II
	Bcl I	A/GATCA	Sau3A I
	Bgl II	A/GATCT	Bgl II, Sau3A I, Xho II
	Sau3A I	A/GATC	Sau3A I
	Xho II	A/GATCY	Bin I[f], Bgl II[f], Sau3A I, Xho II

TABLE II. (Continued)

Ligation of[a]			
Donor end	Acceptor end	Recombined sequence	Recleaved by
Bsp1286[d]	Apa I	GGGCC/C	Apa I, Ban II, Bsp1286, Hae III, Nla IV, Sau96 I
	Ban II[b]	GRGCY/C	Alu I[g], Apa I[g], Ban II, Bsp1286, Hae III[g], HgiA I[g], Nla IV[g], Sau96 I[g], Sst I[g]
	Bsp1286[d]	GDGCH/C	Alu I[j], Apa I[j], Ban II[k], Bsp1286, Hae III[j], HgiA I[k], Nla IV[k], Sau96 I[k], Sst I[k]
	HgiA I[b]	GWGCW/C	Alu I[g], Ban II[g], Bsp1286, HgiA I, Sst I
	Nsi I	GTGCA/T	---
	Pst I	GTGCA/G	---
	Sst I	GAGCT/C	Alu I, Ban II, Bsp1286, HgiA I, Sst I
BspM II	Ava I[b]	T/CCGGG	Bcn I, Cau II, Hpa II, ScrF I
	BspM II	T/CCGGA	BspM II, Hpa II
	Sec I[e]	T/CCGGG	Bcn I, Cau II, Hpa II, ScrF I
	Xma I	T/CCGGG	Bcn I, Cau II, Hpa II, ScrF I
BssH II	Afl III[b]	G/CGCGT	Hha I, HinP I, Tha I
	BssH II	G/CGCGC	BssH II, Hha I, HinP I, Tha I
	Mlu I	G/CGCGT	Hha I, HinP I, Tha I
	Sec I[e]	G/CGCGG	Hha I, HinP I, Tha I
BstE II[b]	BstE II[b]	G/GTNACC	BstE II, Hph I[f], Mae III
	Mae III[b]	G/GTNAC	Hph I[g], Mae III
BstN I[c]	BstN I[c]	CC/WGG	BstN I, EcoR II, ScrF I
	Fnu4H I[b]	CC/WGC	---
	ScrF I[b]	CC/WGG	BstN I, EcoR II, ScrF I
Cau II[c]	Cau II[c]	CC/SGG	Bcn I, Cau II, Hpa II, ScrF I
	Fnu4H I[b]	CC/SGC	Hpa II[f]
	ScrF I[b]	CC/SGG	Bcn I, Cau II, Hpa II, ScrF I
Cfr I	Cfr I	Y/GGCCR	Bal I[g], Cfr I, Gdi II[f], Hae III, Xma III[g]
	Gdi II	Y/GGCCG	Cfr I, Gdi II, Hae III, Xma III[f]
	Not I	Y/GGCCGC	Cfr I, Fnu4H I, Gdi II, Hae III, Xma III[f]
	Xma III	Y/GGCCG	Cfr I, Gdi II, Hae III, Xma III[f]
Cla I	Acc I[b]	AT/CGAC	Taq I
	Aha II	AT/CGYC	---
	Asu II	AT/CGAA	Taq I
	Cla I	AT/CGAT	Cla I, Taq I
	HinP I	AT/CGC	---
	Hpa II	AT/CGG	---
	Mae II	AT/CGT	---
	Nar I	AT/CGCC	---

TABLE II. (Continued)

Ligation of[a]			
Donor end	Acceptor end	Recombined sequence	Recleaved by
Dde I[b]	Dde I[b]	C/TNAG	Dde I
	Esp I[b]	C/TNAGC	Dde I
	Sau I[b]	C/TNAGG	Dde I, Mnl I[g]
Dpn I	Dpn I	G^{m}A/TC	Dpn I, Sau3A
	Nla IV[p]	G^{m}A/TCC	Dpn I, Sau3A
Dra I	Dra I	TTT/AAA	Dra I
	Nru I	TTT/CGA	Taq I
Dra II[b]	Ava II[c]	RG/GWCC	Ava II, Nla IV[f], Sau96 I
	Dra II[b]	RG/GNCCY	Apa I[g], Ava II[f], Ban II[g], Bsp1286[g], Dra II, Hae III[f], Nla IV[f], PpuM I[f], Sau96 I
	PpuM I[c]	RG/GWCCY	Ava II, Dra II, Nla IV[f], PpuM I, Sau96 I
	Rsr II[c]	RG/GWCCG	Ava II, Nla IV[f], Sau96 I
	Sau96 I[c]	RG/GNCC	Apa I[h], Ava II[f], Ban II[h], Bsp1286[h], Hae III[f], Nla IV[f], Sau96 I
EcoR I	EcoR I	G/AATTC	EcoR I
EcoR II[c]	EcoR II[c]	/CCWGG	BstN I, EcoR II, ScrF I
EcoR V	Alu I	GAT/CT	Sau3A I
	Bal I	GAT/CCA	Bin I, Sau3A I
	EcoR V	GAT/ATC	EcoR V
	Hae III	GAT/CC	Bin I, Sau3A I
	Mst I	GAT/GCA	SfaN I
	Nla IV[r]	GAT/BCA	Bin I[i], Hinf I[i], Sau3A I[i], SfaN I[i]
	Nru I	GAT/CGA	Sau3A I, Taq I
	NspB II	GAT/CKG	Sau3A I
	Pvu II	GAT/CTG	Sau3A I
	Stu I	GAT/CCT	Bin I, Sau3A I
	Tha I	GAT/CG	Sau3A I
Esp I[b]	Dde I[b]	GC/TNAG	Dde I
	Esp I[b]	GC/TNAGC	Dde I, Esp I
	Sau I[b]	GC/TNAGG	Dde I, Mnl I[g]
Fnu4H I[b]	BstN I[c]	GC/WGG	---
	Fnu4H I[b]	GC/NGC	Bbv I[f], Fnu4H I
	Cau II[c]	GC/SGG	Hpa II[f]
	ScrF I[b]	GC/NGG	Hpa II[g]
Gdi II	Cfr I	Y/GGCCR	Bal I[g], Cfr I, Gdi II[f], Hae III, Xma III[g]
	Gdi II	Y/GGCCG	Cfr I, Gdi II, Hae III, Xma III[f]
	Not I	Y/GGCCGC	Cfr I, Fnu4H I, Gdi II, Hae III, Xma III[f]
	Xma III	Y/GGCCG	Cfr I, Gdi II, Hae III, Xma III[f]

TABLE II. (Continued)

Ligation of[a]			
Donor end	Acceptor end	Recombined sequence	Recleaved by
Hae II	Hae II	RGCGC/Y	Aha II[g], Ban I[g], Hae II, Hha I, HinP I, Nar I[g], Nla IV[g]
Hae III	Bal I	GG/CCA	Hae III
	EcoR V	GG/ATC	Bin I, Sau3A I
	Hae III	GG/CC	Hae III
	Nla IV	GG/NCC	Ava II[f], Hae III[f], Sau96 I
	Stu I	GG/CCT	Hae III
HgiA I[b]	Ban II[b]	GAGCT/C	Alu I, Ban II, Bsp1286, HgiA I, Sst I
	Bsp1286[d]	GWGCW/C	Alu I[g], Ban II[g], Bsp1286, HgiA I, Sst I[g]
	HgiA I[b]	GWGCW/C	Alu I[g], Ban II[g], Bsp1286, HgiA I, Sst I[g]
	Nsi I	GTGCA/T	---
	Pst I	GTGCA/G	---
	Sst I	GAGCT/C	Alu I, Ban II, Bsp1286, HgiA I, Sst I
Hha I	Hha I	GCG/C	Hha I, HinP I
Hind II	Hind II	GTY/RAC	Acc I[g], Hind II, Hpa I[g], Sal I[g], Taq I[g]
	Hpa I	GTY/AAC	Hind II, Hpa I[f]
	Nla IV[p]	GTY/ACC	Hph I[f], Mae III
	Nru I	GTT/CGA	Taq I
	NspB II[q]	GTC/CGG	Hpa II
	Rsa I	GTY/AC	Mae III
	Sca I	GTY/ACT	Mae III
Hind III	Hind III	A/AGCTT	Alu I, Hind III
Hinf I[b]	Hinf I[b]	G/ANTC	Hinf I
HinP I	Acc I[b]	G/CGAC	---
	Aha II	G/CGYC	Hga I[f], Hha I[f], HinP I[f]
	Asu II	G/CGAA	---
	Cla I	G/CGAT	---
	HinP I	G/CGC	Hha I, HinP I
	Hpa II	G/CGG	---
	Mae II	G/CGT	---
	Nar I	G/CGCC	Hha I, HinP I
	Taq I	G/CGA	---
Hpa I	Hind II	GTT/RAC	Hind II, Hpa I[f]
	Hpa I	GTT/AAC	Hind II, Hpa I
	Nla IV[p]	GTT/ACC	Mae III
	Nru I	GTT/CGA	Taq I
	Rsa I	GTT/AC	Mae III
	Sca I	GTT/ACT	Mae III

TABLE II. (Continued)

Ligation of[a]			
Donor end	Acceptor end	Recombined sequence	Recleaved by
Hpa II	Acc I[b]	C/CGAC	---
	Aha II	C/CGYC	---
	Asu II	C/CGAA	---
	Cla I	C/CGAT	---
	HinP I	C/CGC	---
	Hpa II	C/CGG	Hpa II
	Mae II	C/CGT	---
	Nar I	C/CGCC	---
	Taq I	C/CGA	---
Kpn I	Kpn I	GGTAC/C	Asp718, Ban I, Kpn I, Nla IV, Rsa I
Mae I	Mae I	C/TAG	Mae I
	Nde I	C/TATG	---
Mae II	Acc I[b]	A/CGAC	---
	Aha II	A/CGYC	Mae II[f]
	Asu II	A/CGAA	---
	Cla I	A/CGAT	---
	HinP I	A/CGC	---
	Hpa II	A/CGG	---
	Mae II	A/CGT	Mae II
	Nar I	A/CGCC	---
	Taq I	A/CGA	---
Mae III[b]	BstE II[b]	/GTNACC	Hph I[g], Mae III
	Mae III[b]	/GTNAC	Mae III
Mlu I	Afl III[b]	A/CGCGT	Afl III, Mlu I, Tha I
	BssH II	A/CGCGC	Hha I, HinP I, Tha I
	Mlu I	A/CGCGT	Afl III, Mlu I, Tha I
	Sec I[e]	A/CGCGG	Tha I
Mst I	EcoR V	TGC/ATC	SfaN I
	Mst I	TGC/GCA	Hha I, HinP I, Mst I
	Nae I	TGC/GGC	Fnu4H I
	Nla IV[p]	TGC/GCC	Hha I, HinP I
	NspB II[q]	TGC/CGG	Hpa II
Nae I	Mst I	GCC/GCA	Fnu4H I
	Nae I	GCC/GGC	Hpa II, Nae I
	Nla IV[q]	GCC/KCC	Fnu4H I[f], Mnl I[f]
	NspB II[q]	GCC/CGG	Bcn I, Cau II, Hpa II, ScrF I
	Sma I	GCC/GGG	Bcn I, Cau II, Hpa II, ScrF I

TABLE II. (Continued)

Ligation of[a]			
Donor end	Acceptor end	Recombined sequence	Recleaved by
Nar I	Acc I[b]	GG/CGAC	---
	Aha II	GG/CGYC	Aha II, Ban I[f], Hae II[f], Hga I[f], Hha I[f], HinP I[f], Nar I[f], Nla IV[f]
	Asu II	GG/CGAA	---
	Cla I	GG/CGAT	---
	HinP I	GG/CGC	Hha I, HinP I
	Hpa II	GG/CGG	---
	Mae II	GG/CGT	---
	Nar I	GG/CGCC	Aha II, Ban I, Hae II, Hha I, HinP I, Nar I, Nla IV
	Taq I	GG/CGA	---
Nco I	Afl III[b]	C/CATGT	Nla III
	Nco I	C/CATGG	Nco I, Nla III, Sec I, Sty I
	Sec I[e]	C/CATGG	Nco I, Nla III, Sec I, Sty I
	Sty I[b]	C/CATGG	Nco I, Nla III, Sec I, Sty I
Nde I	Mae I	CA/TAG	---
	Nde I	CA/TATG	Nde I
Nhe I	Avr II	G/CTAGG	Mae I
	Nhe I	G/CTAGC	Mae I, Nhe I
	Sec I[e]	G/CTAGG	Mae I
	Spe I	G/CTAGT	Mae I
	Sty I[b]	G/CTAGG	Mae I
	Xba I	G/CTAGA	Mae I
Nla III	Nla III	CATG/	Nla III
	Nsp I	CATG/Y	Nla III
	Sph I	CATG/C	Nla III
Nla IV	Alu I	GGC/CT	Hae III
	Bal I	GGN/CCA	Ava II[f], Hae III[f], Sau96 I
	EcoR V	GGV/ATC	Bin I[i], Hinf I[i], Sau3A I[i], SfaN I[i]
	Hae III	GGN/CC	Ava II[f], Hae III[f], Sau96 I
	Hind II	GGT/RAC	Hph I[f], Mae III
	Hpa I	GGT/AAC	Mae III
	Mst I	GGC/GCA	Hha I, HinP I
	Nae I	GGM/GGC	Fnu4H I[f], Mnl I[f]
	Nla IV	GGN/NCC	Aha II[m], Apa I[m], Asp718[m], Ava II[g], BamH I[m], Ban I[g], Ban II[m], Bin I[m], Bsp1286[m], Hae II[m], Hae III[n], Hha I[m], HinP I[m], Kpn I[m], Nar I[m], Nla IV, Rsa I[m], Sau3A I[m], Sau96 I[o], Xho II[m]

TABLE II. (Continued)

Ligation of[a]			
Donor end	Acceptor end	Recombined sequence	Recleaved by
Not I	Cfr I	GC/GGCCR	Cfr I, Fnu4H I, Gdi II[f], Hae III, Xma III[f]
	Gdi II	GC/GGCCG	Cfr I, Fnu4H I, Gdi II, Hae III, Xma III
	Not I	GC/GGCCGC	Cfr I, Fnu4H I, Gdi II, Hae III, Not I, Xma III
	Xma III	GC/GGCCG	Cfr I, Fnu4H I, Gdi II, Hae III, Xma III
Nru I	Dra I	TCG/AAA	Taq I
	EcoR V	TCG/ATC	Sau3A I, Taq I
	Hind II[q]	TCG/AAC	Taq I
	Hpa I	TCG/AAC	Taq I
	Nla IV[q]	TCG/RAC	Hae III[f], Taq I
	Nru I	TCG/CGA	Nru I, Tha I
	NspB II[q]	TCG/CGG	Tha I
	Rsa I	TCG/AC	Taq I
	Sca I	TCG/ACT	Taq I
	Ssp I	TCG/ATT	Taq I
	Tha I	TCG/CG	Tha I
Nsi I	Bsp1286[d]	ATGCA/C	---
	HgiA I[b]	ATGCA/C	---
	Nsi I	ATGCA/T	Nsi I
	Pst I	ATGCA/G	---
Nsp I	Nla III	RCATG/	Nla III
	Nsp I	RCATG/Y	Afl III[g], Nla III, Nsp I, Sph I[g]
	Sph I	RCATG/C	Nla III, Nsp I, Sph I[f]
NspB II[q]	Alu I	CAG/CT	Alu I
	EcoR V	CAG/ATC	Sau3A I
	Hind II[q]	CCG/GAC	Hpa II
	Mst I	CCG/GCA	Hpa II
	Nae I	CCG/GGC	Bcn I, Cau II, Hpa II, ScrF I
	Nla IV[p]	CMG/GCC	Hae III, Hpa II[f]
	Nru I	CCG/CGA	Tha I
	NspB II[q]	CGM/CKG	Alu I[g], NspB II, Pvu II[g], Sec I[g], Sst II[g], Tha I[g]
	Pvu II	CMG/CTG	Alu I[g], NspB II, Pvu II[g]
	Sma I	CCG/GGG	Bcn I, Cau II, Hpa II, ScrF I, Sec I
	SnaB I	CCG/GTA	Hpa II
	Tha I	CCG/CG	Tha I
PpuM I[c]	Ava II[c]	RG/GWCC	Ava II, Nla IV, Sau96 I
	Dra II[b]	RG/GWCCY	Ava II, Dra II, Nla IV[l], PpuM I, Sau96 I
	PpuM I[c]	RG/GWCCY	Ava II, Dra II, Nla IV[l], PpuM I, Sau96 I
	Rsr II[c]	RG/GWCCG	Ava II, Nla IV[f], Sau96 I
	Sau96 I[b]	RG/GWCC	Ava II, Nla IV[f], Sau96 I

TABLE II. (Continued)

Ligation of[a]			
Donor end	Acceptor end	Recombined sequence	Recleaved by
Pst I	Bsp1286[d]	CTGCA/C	---
	HgiA I[b]	CTGCA/C	---
	Nsi I	CTGCA/T	---
	Pst I	CTGCA/G	Pst I
Pvu I	Pvu I	CGAT/CG	Pvu I, Sau3A I
Pvu II	Alu I	CAG/CT	Alu I
	EcoR V	CAG/ATC	Sau3A I
	Nla IV[p]	CAG/GCC	Hae III
	NspB II	CAG/CKG	Alu I[f], Nsp B II, Pvu II[f]
	Pvu II	CAG/CTG	Alu I, NspB II, Pvu II
Rsa I	Hind II	GT/RAC	Mae III
	Hpa I	GT/AAC	Mae III
	Nla IV[p]	GT/ACC	Rsa I
	Nru I	GT/CGA	Taq I
	Rsa I	GT/AC	Rsa I
	Sca I	GT/ACT	Rsa I
Rsr II[c]	Ava II[c]	CG/GWCC	Ava II, Sau96 I
	Dra II[b]	CG/GWCCY	Ava II, Nla IV[f], Sau96 I
	PpuM I[c]	CG/GWCCY	Ava II, Nla IV[f], Sau96 I
	Rsr II[c]	CG/GWCCG	Ava II, Rsr II, Sau96 I
	Sau96 I[b]	CG/GWCC	Ava II, Sau96 I
Sal I	Ava I[b]	G/TCGAG	Taq I
	Sal I	G/TCGAC	Acc I, Hind II, Sal I, Taq I
	Xho I	G/TCGAG	Taq I
Sau I[b]	Dde I[b]	CC/TNAG	Dde I, Mnl I[g]
	Esp I[b]	CC/TNAGC	Dde I, Mnl I[g]
	Sau I[b]	CC/TNAGG	Dde I, Mnl I[f], Sau I
Sau3A I	BamH I	/GATCC	Bin I, Sau3A I
	Bcl I	/GATCA	Sau3A I
	Bgl II	/GATCT	Sau3A I
	Sau3A I	/GATC	Sau3A I
	Xho II	/GATCY	Bin I[f], Sau3A I
Sau96 I[b]	Ava II[c]	G/GWCC	Ava II, Sau96 I
	Dra II[b]	G/GNCCY	Apa I[h], Ava II[f], Ban II[h], Bsp1286[h], Hae III[f], Nla III[f], Sau96 I
	PpuM I[c]	G/GWCCY	Ava II, Nla IV[f], Sau96 I
	Rsr II[c]	G/GWCCG	Ava II, Sau96 I
	Sau96 I[b]	G/GNCC	Ava II[f], Hae III[f], Sau96 I
Sca I	Hind II	AGT/RAC	Mae III
	Hpa I	AGT/AAC	Mae III
	Nla IV[p]	AGT/ACC	Rsa I
	Nru I	AGT/CGA	Taq I
	Rsa I	AGT/AC	Rsa I
	Sca I	AGT/ACT	Rsa I, Sca I

TABLE II. Cleavage sequence pairs compatible for ligation.

Ligation of[a]			
Donor end	Acceptor end	Recombined sequence	Recleaved by
ScrF I[b]	BstN I[c]	CC/WGG	BstN I, EcoR II, ScrF I
	Cau II[c]	CC/SGG	Bcn I, Cau II, Hpa II, ScrF I
	Fnu4H I[b]	CC/NGC	Hpa II[g]
	ScrF I[b]	CC/NGG	Bcn I[f], BstN I[f], Cau II[f], EcoR II[f], Hpa II[f], ScrF I
Sec I[e]	Afl III[b]	C/CRYGT	Mae II[g], Nla III[g], Tha I[g]
	Ava I[b]	C/CCGGG	Ava I, Bcn I, Cau II, Hpa II, ScrF I, Sec I, Sma I, Xma I
	Avr II	C/CTAGG	Avr II, Mae I, Sec I, Sty I
	BspM II	C/CCGGA	Bcn I, Cau II, Hpa II, ScrF I
	BssH II	C/CGCGC	Hha I, HinP I, Tha I
	Mlu I	C/CGCGT	Tha I
	Nco I	C/CATGG	Nco I, Nla III, Sec I, Sty I
	Nhe I	C/CTAGC	Mae I
	Sec I[e]	C/CNNGG	Ava I[m], Avr II[m], Bcn I[n], BstN I[g], Cau II[n], EcoR II[g], Hpa II[n], Mae I[m], Mnl I[m], Nco I[m], Nla III[m], NspB II[m], ScrF I[o], Sec I, Sma I[m], Sst II[m], Sty I[g], Tha I[m], Xma I[m]
	Sty I[b]	C/CWWGG	Avr II[g], Mae I[g], Nco I[g], Nla III[g], Sec I, Sty I
	Xba I	C/CTAGA	Mae I
	Xma I	C/CCGGG	Ava I, Bcn I, Cau II, Hpa II, ScrF I, Sec I, Sma I, Xma I
Sma I	Nae I	CCC/GGC	Bcn I, Cau II, Hpa II, ScrF I
	Nla IV[p]	CCC/TCC	Mnl I
	NspB II[q]	CCC/CGG	Bcn I, Cau II, Hpa II, ScrF I, Sec I
	Sma I	CCC/GGG	Ava I, Bcn I, Cau II, Hpa II, ScrF I, Sec I, Sma I, Xma I
SnaB I	NspB II[q]	TAC/CGG	Hpa II
	SnaB I	TAC/GTA	Mae II, SnaB I
Spe I	Avr II	A/CTAGG	Mae I
	Nhe I	A/CTAGC	Mae I
	Sec I[e]	A/CTAGG	Mae I
	Spe I	A/CTAGT	Mae I, Spe I
	Sty I[b]	A/CTAGG	Mae I
	Xba I	A/CTAGA	Mae I
Sph I	Nla III	GCATG/	Nla III
	Nsp I	GCATG/Y	Nla III, Nsp I, Sph I[f]
Ssp I	Nru I	AAT/CGA	Taq I
	Ssp I	AAT/ATT	Ssp I

TABLE II. (Continued)

Ligation of[a]			
Donor end	Acceptor end	Recombined sequence	Recleaved by
Sst I	Ban II[b]	GAGCT/C	Alu I, Ban II, Bsp1286, HgiA I, Sst I
	Bsp1286[d]	GAGCT/C	Alu I, Ban II, Bsp1286, HgiA I, Sst I
	HgiA I[b]	GAGCT/C	Alu I, Ban II, Bsp1286, HgiA I, Sst I
	Sst I	GAGCT/C	Alu I, Ban I, Bsp1286, HgiA I, Sst I
Sst II	Sst II	CCGC/GG	NspB II, Sec I, Sst II, Tha I
Stu I	Bal I	AGG/CCA	Hae III
	EcoR V	AGG/ATC	Bin I, Sau3A I
	Hae III	AGG/CC	Hae III
	Nla IV	AGG/NCC	Ava II[f], Hae III[f], Sau96 I
	Stu I	AGG/CCT	Hae III, Stu I
Sty I[b]	Afl III[b]	C/CATGT	Nla III
	Avr II	C/CTAGG	Avr II, Mae I, Sec I, Sty I
	Nco I	C/CATGG	Nco I, Nla III, Sec I, Sty I
	Nhe I	C/CTAGC	Mae I
	Sec I[e]	C/CWWGG	Avr II[g], Mae I[g], Nco I[g], Nla III[g], Sec I, Sty I
	Spe I[b]	C/CTAGT	Mae I
	Sty I[b]	C/CWWGG	Avr II[g], Mae I[g], Nco I[g], Nla III[g], Sec I, Sty I
	Xba I	C/CTAGA	Mae I
Taq I	Acc I[b]	T/CGAC	Taq I
	Aha II	T/CGYC	---
	Asu II	T/CGAA	Taq I
	Cla I	T/CGAT	Taq I
	HinP I	T/CGC	---
	Hpa II	T/CGG	---
	Mae II	T/CGT	---
	Nar I	T/CGCC	---
	Taq I	T/CGA	Taq I
Tha I	EcoR V	CG/ATC	Sau3A I
	Nla IV[p]	CG/GCC	Hae III
	Nru I	CG/CGA	Tha I
	NspB II[q]	CG/CGG	Tha I
	Tha I	CG/CG	Tha I
Xba I	Avr II	T/CTAGG	Mae I
	Nhe I	T/CTAGC	Mae I
	Sec I[e]	T/CTAGG	Mae I
	Spe I[b]	T/CTAGT	Mae I
	Sty I[b]	T/CTAGG	Mae I
	Xba I	T/CTAGA	Mae I, Xba I
Xho I	Ava I[b]	C/TCGAG	Ava I, Taq I, Xho I
	Sal I	C/TCGAC	Taq I
	Xho I	C/TCGAG	Ava I, Taq I, Xho I

TABLE II. (Continued)

Ligation of[a]			
Donor end	Acceptor end	Recombined Sequence	Recleaved by
Xho II	BamH I	R/GATCC	BamH I[f], Bin I, Nla IV[f], Sau3A I, Xho II
	Bcl I	R/GATCA	Bin I[f], Sau3A I[f]
	Bgl II	R/GATCT	Bgl II[f], Bin I[f], Sau3A I, Xho II[f]
	Sau3A I	R/GATC	Bin I[f], Sau3A I
	Xho II	R/GATCY	BamH I[g], Bgl II[g], Bin I[1], Nla IV[g], Sau3A I, Xho II
Xma I	Ava I[b]	C/CCGGG	Ava I, Bcn I, Cau II, Hpa II, ScrF I, Sec I, Sma I, Xma I
	BspM II	C/CCGGA	Bcn I, Cau II, Hpa II, ScrF I
	Sec I[e]	C/CCGGG	Ava I, Bcn I, Cau II, Hpa II, ScrF I, Sec I, Sma I, Xma I
	Xma I	C/CCGGG	Ava I, Bcn I, Cau II, Hpa II, ScrF I, Sec I, Sma I, Xma I
Xma III	Cfr I	C/GGCCR	Cfr I, Gdi II[f], Hae III, Xma III[f]
	Gdi II	C/GGCCG	Cfr I, Gdi II, Hae III, Xma III
	Not I	C/GGCCGC	Cfr I, Fnu4H I, Gdi II, Hae III, Xma III
	Xma III	C/GGCCG	Cfr I, Gdi II, Hae III, Xma III

a) The enzyme combinations listed are those predicted from their sequence specificity to generate DNA ends which ligate readily, without additional modifications. The "Recombined Sequence" resulting from the 5'-phosphate "Donor End" with the 3'-hydroxyl "Acceptor End" are expected to be "Recleaved by" the enzymes listed in the fourth column. Only those combinations of blunt (flush) end producing enzymes which will provide a cleavable hybrid site are listed in this table. A representative enzyme for each known cleavage sequence is listed, and it is assumed that isoschizomers will function in a similar manner. Those enzymes which cleave the DNA at a location outside of the recognition sequence and are not included here.

Each recombined sequence is written 5' 3' with the "/" indicating the location of ligation of the two DNA ends. Other symbols and the references are as described for Table I.

b)-e) For some enzymes the cleavage sequence is ambiguous due to, for example, purine/pyrimidine "wobbles." Only a fraction of these possible DNA ends can combine with an end created by the second enzyme. The fraction of ends that can combine are indicated by the corresponding superscripts on the enzyme name; fraction, superscript); 1/4, b); 1/2, c); 1/9, d); and 1/16, e).

f)-o) Although expected to ligate together, some enzyme combinations generate recombined sequences that are not recognized by any known restriction endonuclease and, therefore, are not predicted to be recleaved ("---"). Other pairs of restriction enzymes generate a series of end combinations of which only a fraction are recleavable by certain enzymes. For those enzymes that are predicted to cleave only a fraction of the recombined sequences the corresponding superscript appears on the enzyme name; fraction, superscript); 1/2, f); 1/4, g); 1/8, h); 1/3, i); 1/9, j); 4/9, k); 3/4, l); 1/16, m); 3/16, n); and 7/16, o).

p)-r) Some blunt end producing enzymes have a degenerate recognition sequence, for example, a purine/pyrimidine "wobble." Only the fraction of these possible DNA ends that generate cleavable recombined sequences are included in this table. The fraction of ends is indicated by the corresponding superscript on the enzyme name; fraction, superscript); 1/4, p); 1/2, q); and 3/4, r). If no superscript appears, all possible sequences are recleavable.

BUFFER SYSTEMS

A large number of restriction endonucleases have been isolated from a variety of organisms in many different laboratories. This has resulted in the appearance of a large array of reaction buffers for these enzymes. Although restriction endonucleases are unique proteins with individual kinetic optima, they do exibit similar behavior and in some cases their reaction conditions are quite similar. In order to reduce the complexity of the buffer array, consolidating the reaction conditions seems appropriate. However, one must resist the arbitrary application of a universal buffer and base the consolidation on experimentally obtained activity data, so as to avoid unexpected and undesirable results [1].

This laboratory completed a reaction buffer study [14], examining sixty different enzymes for consolidation into four buffer systems. In order for a buffer system to be recommended for a particular enzyme, the performance of the enzyme must have met two criteria. First, the activity was at least 70% of that in the previously accepted buffer and second, no increase in any possible contaminating exo-and non-specific endonucleases was detected. Forty-nine of the enzymes were consolidated, while the remainder were recommended to remain in their original buffer (Table III). Further, to assist in the prediction of activities of multiple enzyme digests the relative activities of each enzyme were determined for the four buffers. In some cases, sequential reactions where the NaCl concentration is adjusted between the two different enzyme additions may be preferable to co-digestion to achieve optimal performance.

TABLE III. Buffer systems[a].

Restriction endonuclease	Recommended buffer system[b]	Relative activity in buffer systems[c]			
		1	2	3	4
Acc I	1	100	< 10	---[d]	100
Alu I	1	100	67	40	80
Apa I	4	25	< 10	< 10	100
Ava I	2	83	100	25	125
Ava II	2	50	100	20	100

TABLE III. (Continued)

Restriction endonuclease	Recommended buffer system [b]	Relative activity in buffer systems [c]			
		1	2	3	4
Bal I	---[e]	30	< 10	< 10	20
BamH I	3	75	125	100	40
Bcl I	2	140	100	60	50
Bgl I	2	20	100	100	40
Bgl II	3	100	100	100	50
BstE II	2	75	100	75	100
Cfo I	1	100	20	< 10	90
Cla I	1	100	100	125	125
Cvn I	4	40	<10	<10	100
Dde I	2	30	100	50	20
Dpn I	4	67	100	67	100
Dra I	1	100 [d]	70	70	135
EcoR I	3	---[d]	120	100	80
EcoR II	---[f]	< 10	< 10	< 10	< 10
EcoR V	2	25	100	125	30
Hae II	2	125	100	67	125
Hae III	2	75	100	80	125
Hha I	2	75	100	50	30
Hinc II	4	< 10	50	50	100
Hind III	2	30	100	50	80
Hinf I	2	100	100	50	80
Hpa I	4	30	50	10	100
Hpa II	---[l]	67	10	---[d]	60
Kpn I	4	60	< 10	< 10	100
Mbo I	2	120	100	60	50
Mbo II	1	100	70	30	85
Mlu I	3	25	50	100	15
Msp I	1	100	25	25	30
Nar I	1	100	30	20	< 10
Nci I	---[l]	< 10	< 10	< 10	80
Nco I	3	100	133	100	60
Nde I	2	10	100	100	15
Nde II	---[m]	50	75	133	10
Nru I	---[g]	< 10	< 10	67	15
Nsi I	3	50	125	100	20
Pst I	2	35	100	50	50
Pvu I	---[g]	20	33	60	10
Pvu II	---[h]	30	50	30	40
Rsa I	1	100	50	10	80
Sal I	3	<10	30	100	<10
Sau3A I	4	<10	<10	<10	100
Sau96 I	---[i]	20 [d]	10 [d]	< 10 [d]	< 10
Sca I	---[h]	---[k]	---[k]	---[k]	< 10
Sma I	4	---[k]	---[k]	---[k]	100
Sph I	---[h]	150	100	40	40
Sst I	2	150	100	50	90
Sst II	2	135	100	<10	70
Stu I	2	120	100	50	100
Sty I	3	50	100	100	< 10
Taq I	2	50	100	75	50
Tha I	1	100	67	< 10	100
Xba I	2	60	100	30	125
Xho I	2	50	100	100	50
Xma III	---[j]	50	< 10	< 10	20
Xor II	4	30	< 10	< 10	100

a) The data were kindly provided by C. Sampson, A. Kerlin, and A. Marschel of this laboratory [14].

b) Of those tested this buffer demonstrated the highest acceptable ratio of restriction endonuclease to contaminating non-specific endo- and exonuclease activities. Where indicated ("---") the highest acceptable ratio was observed only in the special buffer.

c) Enzyme activity was measured in each buffer system by the standard unit assay procedure. The values (units/μl enzyme) were normalized to the activity in the recommended buffer, which was given the value '100'. The buffers were; 1, 50 mM Tris-HCl (pH 8.0), 10 mM $MgCl_2$; 2, 50 mM Tris-HCl (pH 8.0), 10 mM $MgCl_2$, 50 mM NaCl; 3, 50 mM Tris-HCl (pH 8.0), 10 mM $MgCl_2$, 100 mM NaCl; and 4, 20 mM Tris-HCl (pH 7.4), 5 mM $MgCl_2$, 50 mM KCl.

d) This buffer causes significant 'star activity' for this enzyme and should not be used.

e) 50 mM Tris-HCl (pH 8.5), 5 mM $MgCl_2$

f) 50 mM Tris-HCl (pH 7.4), 6 mM $MgCl_2$, 50 mM NaCl, 50 mM KCl, 1 mM dithiothreitol

g) 50 mM Tris-HCl (pH 8.0), 10 mM $MgCl_2$, 50 mM NaCl, 50 mM KCl

h) 50 mM Tris-HCl (pH 7.4), 6 mM $MgCl_2$, 50 mM NaCl, 50 mM KCl

i) 10 mM Tris-HCl (pH 9.0), 12 mM $MgCl_2$, 100 mM KCl

j) 10 mM Tris-HCl (pH 8.2), 8 mM $MgCl_2$

k) <u>Sma</u> I requires K^+ for activity and exhibits very low activity with Na^+ containing buffers.

l) 20 mM Tris-HCl (pH 7.4), 10 mM $MgCl_2$

m) 100 mM Tris-HCl (pH 7.6), 10 mM $MgCl_2$, 150 mM NaCl, 1 mM dithiothreitol.

HEAT SENSITIVITY

The assay temperature for most restriction endonucleases is 37°C. The effect of elevated reaction temperature on restriction endonuclease activity is variable. Above 42°C, <u>EcoR</u> I [15] and <u>Hae</u> II [16] were inactivated, while at 70°C [16] <u>Hae</u> III was fully active. And, of course, those enzymes from thermophilic organisms, e.g., <u>Tha</u> I and <u>Taq</u> I were active at 60°C. After completing a reaction, inactivation of the restriction endonuclease by incubation for 10 minutes at 65°C (Table IV) was only possible in a number of instances [17]. Thus, a phenol extraction was required in order to properly stop cleavage by all of the

enzymes. This information was especially useful if a ligation reaction
was planned immediately to follow the digestion.

In order to determine enzyme stability at normal incubation temperatures,
a second study was performed in this laboratory [18]. Restriction
endonucleases were incubated under normal reaction conditions (buffer,
dilution and temperature), but without substrate (DNA). After various
times, DNA was added and cleavage activity was assayed (Table IV). Over
25% of the enzymes lost some activity, while another 33% lost all activity
within the first hour of pre-incubation without DNA. On the other hand, 24%
of the enzymes retained full activity for four hours. (Note, all enzymes
were fully active in the absence of any pre-incubation) In the presence of
substrate (DNA), the stability improved for a number of the enzymes that
earlier were shown to be unstable. The data in this table suggests a
relative stability of the various restriction endonucleases to incubation
conditions. This will be useful, for example in the planning of extended
incubations for the cleavage of large amounts of DNA.

TABLE IV. Enzyme stability.

Enzyme	Hours of pre-incubation with retained activity[a]		Activity after incubation at 65°C, 10 minutes[b]
	DNA absent	DNA present	
Acc I	4		++
Alu I	4		-
Ava I	4		+
Ava II	2 (4)		-
Bal I	0	4	-
BamH I	0	1	++
Bcl I	4		++
Bgl I	0	4	-
Bgl II	0 (4)		++
BstE II	0 (2)		++
Cfo I	0	0	+
Cla I	4		+
Cvn I	0		-
Dde I	0 (2)		++
Dpn I	0 (4)	4	-
Dra I	0	2	-
EcoR I	4		+
EcoR II	0	1	-
EcoR V	0 (1)		
Hae II	0 (4)		-
Hae III	1 (4)		++
Hha I	0 (2)		-
Hinc II	0 (4)		-
Hind III	1 (4)		++
Hinf I	0 (3)		++
Hpa I	0 (4)		++
Hpa II	4		++
Kpn I	0 (1)	1	-
Mbo I	0 (4)		-
Mbo II	2 (4)		+
Msp I	0	1	-
Nar I	4		-
Nci I	0 (3)		+

TABLE IV. (Continued)

Enzyme	Hours of pre-incubation with retained activity[a]		Activity after incubation at 65°C, 10 minutes[b]
	DNA absent	DNA present	
Nde I	0 (4)		++
Nde II	0 (4)		-
Nru I	0 (2)	4	+
Pst I	0	1	+
Pvu I	2 (4)		++
Pvu II	0	3	+
Rsa I	0	1	-
Sal I	1 (4)		++
Sau3A I	0	1	-
Sau96 I	4		++
Sma I	0	3	+
Sph I	0		-
Sst I	4		-
Sst II	4		++
Taq I	2 (4)		++
Tha I	0 (1)		++
Xba I	1 (4)		++
Xho I	4		++
Xma III	4		+
Xor II	0 (1)	4	++

[a] The data were kindly provided by J. Crouse and D. Amorese [18]. Enzyme (1 Unit) was pre-incubated in the appropriate reaction buffer (Table III) in the presence or absence of sheared salmon sperm DNA. After 0, 1, 2, 3 or 4 hours 1 μg of substrate DNA was added, the reaction incubated for another hour, and the products analyzed by agarose gel electrophoresis. Full enzyme activity was observed after the indicated hours of pre-incubation in the absence of DNA. In some of the cases where activity diminished before 4 hours of pre-incubation the figure in parenthesis indicated the hours of pre-incubation that at least partial activity was observed. In the cases where no enzyme activity was observed upon pre-incubation (0), the stabilizing effect of DNA was tested. The figures indicate the hours of pre-incubation in the presence of DNA that at least partial enzyme activity was observed.

[b] Reactions were performed in a 100 μl volume with 2 μg of DNA and 20 or 40 units of enzyme for 60 minutes at the appropriate temperature. The reactions were then incubated at 65°C for 10 minutes and placed on ice. To a 50 μl aliquot, 1 μg of DNA was added and the reaction incubated for an additional 5 hours at the appropriate temperature. A "-" indicates no activity, a "+" indicates partial activity and a "++" indicates complete activity was detected in the second cleavage reaction [17, 27].

FACTORS INFLUENCING ACTIVITY

Even though a restriction endonuclease reaction is composed of relatively few components a number of factors can influence activity [1]. Enzyme activity can be inhibited by impurities inadvertently added to the reaction volume as well as by the environmental conditions imposed. Unexpected reaction products may result from contaminating DNAs or the presence of additional enzyme activities. Cleavage inhibition on the other hand, can be diagnostic of an unusual structure or a base modification in the DNA substrate. The following table (Table V) summarizes a number of factors that influence both positively and negatively the results of a restriction endonuclease reaction. The factors are grouped as involving the DNA substrate, Buffer solution, Enzymic protein, Temperature, Time/Volume, and Additional reagents. With each factor are some notes giving possible explanations for the effects observed and, in some cases, suggesting solutions for problem situations. In other cases, alterations are suggested for increasing the utility of the reaction for certain applications.

TABLE V. Factors influencing restriction endonuclease activity.

Factor	Effect	Notes
1. DNA substrate		
a. Purity	Cleavage inhibition or specificity change caused by contaminants:	
	Tightly associated molecules (e.g., basic proteins, spermine, ethidium bromide[19], distamycin [20])	Before reaction, remove these by chloroform/phenol extraction and/or ethanol precipitation from high salt (2.5 M ammonium acetate). See "Additional reagents". Bound proteins may alter electrophoretic mobility and, thus, expected size of DNA fragments. Incubate in 0.1% SDS at 65 C before gel application.
	Organic solvents (e.g., chloroform, phenol, DMSO) and ionic detergents (e.g., SDS).	Remove by ethanol precipitation from high salt (e.g., 2.5 M ammonium acetate).
	Other nucleic acids	RNA reduces effective enzyme concentration. DNA generates unwanted fragments. Remove before reaction by gel electrophoresis, or by exclusion or ion exchange chromatography.
	Solvent (e.g., NaCl, heavy metals, chelators, 7 > pH > 8)	Remove by ethanol precipitation from 2.5 M ammonium acetate, then lyophilization. Resuspend in low salt DNA buffer (e.g., 20 mM Tris-HCl, pH 7.2, 5 mM NaCl, 0.1 mM Na_2 EDTA). See "Buffer solution".
	Acrylamide/agarose	Presence frequently causes enzyme inhibition. Remove before reaction by ion exchange chromatography.

Table V. (Continued)

Factor	Effect	Notes
b. Structure	Reduced apparent enzyme activity relative to linear duplex DNA by:	
	Supercoiling (e.g., pBR322)	To promote activity, first treat DNA with topoisomerase or linearize with a second restriction endonuclease [1].
	Annealed ends (e.g., left and right ends of lambda DNA)	Heat completed reactions for 10 min at 65 C and quick cool on ice prior to gel electrophoresis.
	Z-DNA	Limits of structural perturbations were measured [21].
	Apparent cleavage of single-stranded DNA (e.g., ØX174(+), M13(+))	Duplex regions within some single-stranded DNAs were cleaved by some enzymes added in excess [16, 22].
c. Sequence	Reduced or no cleavage activity:	
	Absence of recognition sequence	Expected cleavage sequence unexpectedly modified by in vivo mutation or modification (see below), or by imprecise in vitro manipulation (e.g., exonucleolytic removal of terminal nucleotides before ligation).
	Neighboring nucleotides	Cleavage rate influenced up to 10-fold by sequence bordering recognition sequence [23-25].
	End proximity	Hpa II and Mno I require at least one nucleotide 5' to the recognition sequence for cleavage [26].

TABLE V. (Continued)

Factor	Effect	Notes
	Site density	No influence by site density is expected in a standard (enzyme saturating) reaction. Experiments have failed to demonstrate a correlation [27].
d. Modification	Reduced or no cleavage activity: Methylation	Methylation of appropriate nucleotides within the recognition sequence will prevent cleavage by certain enzymes. For complete cleavage use a methylation insensitive isochizomer, or isolate DNA from a methylase deficient source. Methylation of the recognition sequence is required for cleavage by some enzymes (e.g., Dpn I).
	Partial methylation	A subset of cleavage sites in DNAs isolated from E. coli (dam⁺) may be methylated resulting in a persistant, incomplete cleavage pattern [28]. See "Methylation".
	Overlapping methylation	Some recognition sequences may overlap methylated sites preventing cleavage (e.g., for Xba I of sequence TCTAGA with G$\overset{m}{A}$TC). Not all occurances of the sequence in a given DNA will overlap a methylated site resulting in a partial cleavage pattern. 'New' cleavage specificities are obtained by intentional methylation of DNA prior to cleavage [29, 30].

TABLE V. (Continued)

Factor	Effect	Notes
	Other (e.g., 5-hydroxymethyl cytosine, glucosylation)	Activity was influenced by some phage specific DNA modifications [31]. Phosphorothioate – modified internucleolytic linkages at a restriction site inhibits cleavage and can be used to prepare nicked DNA [32].
2. Buffer solution		
a. Buffer	Alteration of cleavage	Tris-HCl is most commonly used, but note pH changes with temperature and concentration. Phosphate buffers carried over to ligase or kinase reactions can be inhibitory. Ionic strength should be sufficient to hold pH in reaction.
b. pH	Alteration of cleavage rate or specificity	Most enzymes are active near pH 7.5, with a broad optimum (where tested). EcoR I alters cleavage specificity ('star activity') when increased from pH 7.2 to 8.5 [33].
c. Me^{++}	No cleavage if absent	Mg^{++} is usual divalent cation, but Mn^{++} can substitute in some cases with reduced activity or specificity change [34]. Other ions (e.g., Cu^{++}, Zn and Co) may not substitute [16]. Stop cleavage by chelation of Mg^{++} with EDTA.
d. Me^{+}	Alteration of cleavage rate	NaCl is typically used from 0 to 100 mM, and can influence activity several fold (see Table III). KCl is required for Sma I activity.

TABLE V. (Continued)

Factor	Effect	Notes
	Alteration of cleavage specificity	Little or no ionic strength resulting from monovalent cations increases altered cleavage specificity ('star activity') for some enzymes.
e. Sulfhydral reagents	Little or no influence	A number of enzymes are insensitive to p-chloromercuribenzoate and N-ethylmaleimide [35]. In general addition of 2-mercapto-ethanol or dithiothreitol to a reaction is usually not required.
3. Enzymic protein		
a. Dilution	Activity of dilute solutions is less stable.	Concentrated enzyme is generally more stable. See also "Altered specificity".
b. Cognate enzyme	Additional cleavage products	Many enzymes are purified from bacteria containing multiple restriction endonucleases (e.g., Haemophilus aegyptius; Hae II and Hae III). Detect by cleavage of DNA of known sequence (e.g., pBR322).
c. Altered specificity	Usually an increase in number and decrease in size of product fragments	Results of 'star activity' observed with several enzymes under high pH, low ionic strength, added glycerol or very high enzyme concentration reaction conditions. [33, 34, 37]
d. Second enzyme added	Possible incomplete combined digest	Sequential restriction reactions can alleviate interference observed by simultaneous double digests.

TABLE V. (Continued)

Factor	Effect	Notes
e. Other nucleases	Inhomogeneity of fragment length or end sequence	Contaminating non-specific endo- and exo-nucleases alter the specific cleavage products of restriction enzyme digests. Reduce problem by using high quality enzymes, pure DNAs and sterile buffers, tubes and pipettes.
f. Protease	Rapid loss of activity	Addition of nuclease-free BSA to enzyme stock dilutions and to reactions reduces rate of activity loss. Also, EGTA or phenylmethyl-sulfonyl chloride have increased the stability of some enzymes.
4. Temperature		
a. Incubation low	Cleavage proceeds at a reduced rate	Incubate enzymes at appropriate temperature, usually 37 C. Enzymes from thermophiles (e.g., Taq I, Tha I) are frequently most active near 60 C. Enzyme should be added last to the reaction. Even on ice (4 C) cleavage occurs.
b. Incubation high	Cleavage inhibited, unaffected, or stimulated	Some enzymes are rapidly inactivated above 42 C (e.g., EcoR I [15] and Hae II [16]). Other enzymes are more active above 37 C (e.g., Hae III [16], Acc I and Ava I [38], and Taq I and Tha I). Some enzymes are stable to incubation for 10 min at 65 C (See Table IV).
c. Storage method	Loss of activity	Solution should be cold, but not frozen; usually -20 C for those containing glycerol, +4 C for those without.

TABLE V. (Continued)

Factor	Effect	Notes
5. Time/volume		
a. Stability	Variable activity loss	Activity versus incubation time usually is not linear, i.e., one unit cleaves 1 g of DNA in one hour, but not necessarily 2 g in two. Variable stability with and without DNA present is also seen. See Table IV.
b. Dilution	Activity may vary with dilution; altered cleavage specificity.	Total reaction volume can influence enzyme stability, uniformity of mixture, viscosity of solution and total protein concentration. Enzyme and glycerol concentration can induce 'star activity', see below.
c. Dispensing	Variable activity	Large errors resulting from pipetting of small volumes of vicous enzyme solutions can lead to irreproducible results.
6. Additional reagents		
a. Enzymes	Altered cleavage pattern	The presence of a second restriction endonuclease may prevent complete digestion. Other nucleic acid binding enzymes may interfere or compete against completion of digestion. A single homogeneous reaction is recommended. See "Enzymic protein". Pre-treatment with DNA methylases can be used to select a subset of sites for cleavage [29, 30].

TABLE V. (Continued)

Factor	Effect	Notes
b. Glycerol	Altered cleavage pattern	The presence of glycerol, especially > 5% (v/v), may induce a less stringent cleavage specificity ('star activity') in certain enzymes. See "Enzymic protein".
c. DNA ligands	Inhibition of selective sites	A subset of cleavage sites within a DNA may be protected from cleavage [19, 20, 36].

REFERENCES

1. R. Fuchs and R. Blakesley in: Methods in Enzymology, Vol. 100: Recombinant DNA, Part B, R. Wu, L. Grossman and K. Moldave, eds. (Academic Press, New York 1983) pp. 3-38.
2. R.D. Wells, R.D. Klein and C.K. Singleton in: The Enzymes, 3rd ed., Vol. 14, Part A, P.D. Boyer, ed. (Academic Press, New York 1981) pp. 157-191.
3. R.J. Roberts, this volume.
4. R. Blakesley in: Gene Amplification and Analysis, Vol 2: Structural Analysis of Nucleic Acids, J.G. Chirikjian and T.S. Papas, eds. (Elsevier/North-Holland, Amsterdam 1981) pp. 85-133.
5. C. Kessler, P.S. Neumaier and W. Wolf, Gene 33, 1-102 (1985).
6. R.J. Roberts, Nucleic Acids Res. 13, r165-r200 (1985).
7. M. McClelland and M. Nelson, Nucleic Acids Res. 13, r201-r207 (1985).
8. R. Calleja, N.T. deMarsac, T. Coursin, H. van Ormondt and A. de Waard, Nucleic Acids Res. 13, 6745-6751 (1985).
9. V. Butkus, S. Klimasaukas, D. Kersulyte, D. Vaitkevicius, A. Lebionka and A. Janulaitas, Nucleic Acids Res. 13, 5727-5746 (1985).
10. K. Mise and K. Nakajima, Gene 36, 363-367 (1985).
11. K. Mise and K. Nakajima, Gene 33, 357-361 (1985).
12. K. Kita, N. Hiraoka, A. Oshima, S. Kadonishi and A. Obayashi, Nucleic Acids Res. 13, 8685-8694 (1985).
13. A. Cornish-Bowden, Nucleic Acids Res. 13, 3021-3030 (1985).
14. C. Sampson, A. Kerlin and A. MarSchel, BRL FOCUS 8 (1986).
15. P.J. Greene, M.S. Poonian, A.L. Nussbaum, L. Tobias, D.E. Garfin, H.W. Boyer and H.M. Goodman, J. Mol. Biol. 99, 237-261 (1975).
16. R.W. Blakesley, J.B. Dodgson, I.F. Nes and R.D. Wells, J. Biol. Chem. 252, 7300-7306 (1977).
17. BRL FOCUS 5:2, 13 (1983).
18. J. Crouse and D. Amorese, BRL FOCUS 8 (1986).
19. M. Osterlund, H. Luthman, S.V. Nilsson and G. Magnusson, Gene 20, 121-125 (1982).
20. V.V. Nosikov, E.A. Braga, A.V. Karlishev, A.L. Zhuze and O.L. Polyanovsky, Nucleic Acids Res. 3, 2293-2301 (1976).
21. F. Azorin, R. Hahn and A. Rich, Proc. Nat. Acad. Sci. 91, 5714-5718 (1984).
22. K. Nishigaki, Y. Kaneko, H. Wakuda, Y. Husimi and T. Tanaka, Nucleic Acids Res. 13, 5747-5760 (1985).
23. M. Thomas and R.W. Davis, J. Mol. Biol. 91, 315-328 (1975).
24. K. Armstrong and W.R. Bauer, Nucleic Acids Res. 10, 993-1007 (1982).

25. H.R. Drew and A.A. Travers, Nucleic Acids Res. 13, 4445-4467 (1985).
26. B.R. Baumstark, R.J. Roberts and U.L. RajBhandary, J. Biol. Chem. 254, 8943-8950 (1979).
27. Unpublished opservations, this laboratory.
28. B. Drieseikelmann, R. Eichenlaub and W. Wackernagel, Biochim. Biophys. Acta 562, 418-428 (1979).
29. M. McClelland, L.G. Kessler and M. Bittner, Proc. Nat. Acad. Sci. 81, 983-987 (1984).
30. M. Nelson, C. Christ and I. Schildkraut, Nucleic Acids Res. 12, 5165-5173 (1984).
31. L.-H. Huang, C.M. Farnet, K.C. Ehrlich and M. Ehrlich, Nucleic Acids Res. 10, 1579-1591 (1982).
32. J.W. Taylor, W. Schmidt, R. Cosstick, A. Okruszek and F. Eckstein, Nucleic Acids Res. 13, 8749-8764 (1985).
33. B. Polisky, P. Greene, D.E. Garfin, B.J. McCarthy, H.M. Goodman and H.W. Boyer, Proc. Nat. Acad. Sci. 72, 3310-3314 (1975).
34. M. Hsu and P. Berg, Biochemistry 17, 131-138 (1978).
35. K. Nath, Arch. Biochem. Biophys. 212, 611-617 (1981).
36. J. Kania and T.G. Fanning, Eur. J. Biochem. 67, 367-371 (1976).

3

Mechanism of Specific Site Location and DNA Cleavage by EcoR I Endonuclease

Brian J. Terry, William E. Jack, and Paul Modrich

Department of Biochemistry
Duke University Medical Center
Durham, North Carolina 27710

I. INTRODUCTION

The EcoR I restriction and modification system (Table I) has become one of the most convenient systems in which to study protein-DNA interactions [1]. Both EcoR I endonuclease and methylase recognize a two-fold symmetrical DNA sequence. The endonuclease cleaves at sites indicated by arrows while the central adenines are methylated on the 6-amino group by the methylase (*) [2, 3], with the latter modification rendering the EcoR I sequence resistant to cleavage by the endonuclease [3]. Like all sequence specific proteins, EcoR I endonuclease displays finite affinity for nonspecific DNA sequences. In this context, the problem of specific recognition can be divided into two questions: (i) What is the mechanistic basis for discrimination of specific and nonspecific sites in thermodynamic and molecular terms? (ii) What role do nonspecific interactions play in the kinetic path(s) utilized for specific site location? In this paper we will summarize work on the thermodynamics of specific and nonspecific interactions and will discuss evidence indicating that nonspecific interactions are involved in the pathways by which EcoR I endonuclease locates an EcoR I sequence and leaves a cleaved site. For discussion of molecular details of endonuclease·DNA interactions, see chapter by Rosenberg et al., this volume.

TABLE I. Components of the EcoR I restriction and modification system.

Component	Active Form	Comment
5'-G↓A-A-T-T-C 3'-C-T-T-A-A↑G (* above first A, * below last A)	Duplex	Single strand not recognized [4 - 6]
Endonuclease	Dimer	DNA-endonuclease complex two-fold symmetric [7, 8]
Methylase	Monomer	DNA-methylase complex probably asymmetric [5]

II. SPECIFICITY OF BINDING AND CLEAVAGE

The EcoR I recognition site would be expected to occur on average once every 4096 bp. The remainder of the DNA will be composed of nonspecific DNA sequences, which may share certain structural features with the recognition site. For example, for each EcoR I site, on the average one would expect eighteen sites which differ from the EcoR I sequence by one base pair. Yet, EcoR I endonuclease cleaves DNA at the recognition site with a high degree of fidelity [9].

EcoR I endonuclease binds with high affinity to EcoR I sites in the absence of Mg^{2+}, conditions under which DNA hydrolysis cannot occur [9 - 13]. Endonuclease binding to the single EcoR I site of the plasmid pBR322 is extremely tight (Table II and ref. [14]). Site specific complexes are readily quantitated by the nitrocellulose filter binding assay [9 - 11, 14, 15] or by preferential cleavage of preformed endonuclease DNA complexes upon the subsequent addition of Mg^{2+} [14]. Both methods yield identical results indicating that the nitrocellulose filter binding method provides an accurate measure of specific complexes.

However, attempts to directly measure nonspecific endonuclease binding by the nitrocellulose filter binding method have proven difficult except at low ionic strength. Consequently, nonspecific binding has generally been measured by competition methods. Such analysis indicates that pBR322ΔRI (in which the EcoR I site has been destroyed) acts as a competitive inhibitor of endonuclease binding to radiolabeled pBR322 [14]. The affinity of EcoR I endonuclease for the pBR322ΔRI molecule is 60-fold less than the affinity for the specific site of pBR322 on a molar basis (Table II). Since pBR322ΔRI does not contain any EcoR I sites, the endonuclease must be binding to nonspecific DNA sequences. If one considers that there are 4367 potential nonspecific binding sites [16] on pBR322ΔRI compared to a single EcoR I site on pBR322, the relative affinity of specific to nonspecific sites is 10^5-fold. Interestingly, pBR322 where the EcoR I site has been modified by EcoR I methylase binds endonuclease with an affinity similar to that of pBR322ΔRI. This indicates that the inability of EcoR I endonuclease to cleave the EcoR I methylated recognition site is at least partially due to the reduced affinity of the endonuclease for the modified EcoR I site.

Although absolute magnitudes differ, comparison of endonuclease binding to lambda DNA shows a similar effect [9, 17]. Binding of EcoR I endonuclease to either of two lambda derivatives containing a single EcoR I site is similar (Table II and [9]). Endonuclease affinity for a lambda derivative lacking EcoR I sites (λ395) is reduced at least 10-fold. Considering λ395 DNA contains 4×10^4 potential nonspecific binding sites, the relative affinity appears to be greater than 10^5-fold [9, 17].

TABLE II. Equilibrium binding constants for EcoR I endonuclease.

Substrate[a]	K_{diss}, M[b]	Buffer[c]	Reference
Contain EcoR I site:			
pBR322	5×10^{-12}	A	14
34 bp DNA (from pBR322)	10×10^{-12}	A	14
d(pCGCGAATTCGCG)	15×10^{-12}	B	15
λ401, λ416	5×10^{-10}	C	9
No EcoR I sites:			
pBR322ΔRI	3×10^{-10} (1×10^{-6})	A	14
Me$_2$pBR322	4×10^{-10} (2×10^{-6})	A	14
λ395	6×10^{-9} (2×10^{-4})	C	9

[a] DNA substrates: 34 bp DNA fragment contains the unique EcoR I pBR322; duplex form of d(CGCGAATTCGCG); pBR322ΔRI has no EcoR I site; Me$_2$pBR322 has been modified with EcoR I methylase; λ401 and λ416 are bacteriophage λ derivatives containing one EcoR I site (sRI 2 and sRI 5, respectively [18]); λ395 contains no EcoR I sites.

[b] Equilibrium dissociation constants are expressed in terms of DNA molecules. Values in parentheses are expressed in terms of potential nonspecific binding sites.

[c] Buffer conditions: A, 0.1 M TrisHCl (pH 7.6), 1 mM EDTA, and 0.05 mg/ml bovine serum albumin at 37°C; B, 0.01 M bis-tris propane (pH 7.4), 0.075 M KCl, 0.1 mM EDTA, 0.05 mM dithiothreitol, and 0.1 mg/ml bovine serum albumin at 24°C; C, 0.05 M Tes (pH 7.0), 0.1 M NaCl, and 1 mM EDTA at 22°C.

The relative affinities for pBR322 and pBR322ΔRI indicate that the affinity of endonuclease for pBR322 is largely due to interaction with the EcoR I site. Confirming this idea, EcoR I endonuclease exhibits similar affinity for a 34-bp fragment isolated from pBR322 which contains the EcoR I site and for the parent 4363-bp DNA molecule. Binding studies with a dodecamer d(pCGCGAATTCGCG) (duplex form) produce an equilibrium dissociation constant of 15 pM at a similar ionic strength (although at a lower temperature) [15]. Therefore, the thermodynamic stability of specific complexes is largely independent of nonspecific DNA sequences.

Thus, EcoR I endonuclease exhibits a large preference for binding to an EcoR I site as compared to nonspecific DNA sequences [14, 15]. This differential binding affinity is partly responsible for the high degree of fidelity of cleavage shown by EcoR I endonuclease; under optimal cleavage conditions, cleavage at an EcoR I recognition site occurs at least 10^7 times more frequently than cleavage at any nonspecific site [9]. However, solution conditions may be chosen to promote cleavage at non-canonical EcoR I sites EcoR I* sites) which is visualized as a reduction in specificity. The basis of this effect is not currently understood.

III. MECHANISM OF SITE LOCATION

The ability of DNA binding proteins to locate their recognition sites by kinetically efficient mechanisms has spawned several models for facilitated diffusion [19, 20]. Facilitated diffusion involves two steps: a protein binds to nonspecific DNA sequences followed by transfer to other sites on the DNA molecule by a dimension-limited or volume-limited diffusion process. von Hippel and colleagues have proposed four general mechanisms for facilitated diffusion along a DNA molecule [20]. The protein may slide along the DNA molecule during transfer, maintaining constant contact with the helix. Alternatively, the protein may undergo microscopic dissociation-reassociation, or hop along the DNA. In the latter mechanism, the protein undergoes microscopic dissociation but reassociates with the helix due to its proximity to the DNA. Both of these transfer mechanisms are positionally correlated since transfer occurs between sites which are nearby as measured along the DNA contour [20].

Two positionally uncorrelated transfer mechanisms have also been proposed [20]. These mechanisms involve transfer among DNA binding sites which are well separated along the DNA contour. Intradomain dissociation-reassociation involves macroscopic dissociation of the protein from the DNA followed by reassociation with another site on the DNA. Since the protein is unconstrained in solution, the probability of reassociating with the original binding site is no greater than binding to any other site on the same DNA molecule. Intersegment transfer may occur when a protein·DNA complex is attacked by a distal DNA site to yield an intermediate where the protein is transiently associated with two DNA segments. Following the decay of this intermediate, the protein remains associated with either the original binding site or the distal DNA segment. Both intradomain transfer and intersegment transfer depend on flexibility of DNA chains to bring potential binding sites into proximity.

Rate of dissociation of specific complexes

Facilitated diffusion models predict association rate enhancement with increasing DNA chain length [19, 20]. Similarly, dissociation rates are expected to increase for longer DNA molecules because they have a greater pool of nonspecific DNA sequences from which dissociation may occur. Analysis of dissociation rate with respect to DNA chain length in the absence of Mg^{2+} indicates that the dissociation rate of specific EcoR I complexes increases with DNA chain length, a finding consistent with a facilitated diffusion mechanism (Fig. 1 and [21]). The lower dissociation rates for short DNA molecules is not due to an affinity for DNA termini since dissociation of covalently closed circular DNA parallels that of linear DNA of the same size. Moreover, the continuous nature of the relationship below 300 bp indicates that translocation models dependent on DNA flexibility play little, if any, role in search kinetics in the

experiments shown in Fig. 1. In fact, the solid curve in Fig. 1 is a fit to
the theoretical equations for the sliding mechanism proposed by von Hippel
[20] using the specific and nonspecific binding constants determined for
EcoR I endonuclease (Table II). Such analysis suggests the endonuclease may
search about 1300 bp per binding event in the absence of Mg^{2+} at an ionic
strength of 0.079 [21].

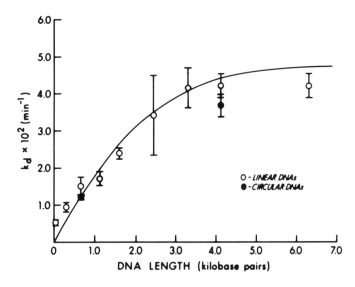

Fig. 1. Dependence of dissociation rate constant on DNA chain length.
Dissociation rate constants were determined under standard binding
conditions: 0.1 M TrisHCl (pH 7.6), 1 mM EDTA, and 0.05 mg/ml bovine serum
albumin at 37°C [21].

Preferential cleavage of longer DNA molecules

Association and dissociation rate constants for site-specific binding
of EcoR I endonuclease are DNA chain length dependent (Fig. 1 and [21]).
Competition cleavage experiments have been performed to determine if DNA
cleavage rates are also chain length dependent [21, 22]. Equimolar mixtures
of pBR322 (4363 bp) and a 34 bp fragment derived from pBR322 containing the
EcoR I site in a central location were cleaved with EcoR I endonuclease
(Table IV). There is a strong preference for the longer DNA at low ionic
strength indicating longer DNA molecules have a kinetic advantage even in
the presence of Mg^{2+} (5 mM). This preferential cleavage decreases with
increasing ionic strength consistent with the reduction in nonspecific DNA
binding if such DNA sequences facilitate site location.

TABLE IV. Preferential Cleavage of Longer DNA Molecules[a]

[NaCl], M	Ionic Strength	Ratio of v_o, 4363bp/34bp
0.025	0.059	12
0.10	0.13	2.0
0.20	0.23	0.94

[a] Reactions contained an equimolar mixture (1 nM) of linear 5'-radiolabeled pBR322 (4363 bp) and 34 bp DNA molecules in 0.02 M TrisHCl (pH 7.6), 0.2 mM EDTA, 5 mM $MgCl_2$, 0.05 mg/ml bovine serum albumin, and NaCl as indicated. Reactions were initiated by addition of EcoR I endonuclease (0.01 nM) and initial rates (v_o) determined for the first 10% of the reaction.

Processive action of EcoR I endonuclease

An alternative method of analyzing the role of nonspecific DNA sequences is to determine whether EcoR I endonuclease cleaves in a processive manner. The efficiency of cleavage at a second EcoR I site could discriminate between positionally correlated and uncorrelated transfer mechanisms. To minimize the effect neighboring DNA sequences may have on cleavage of EcoR I sites [23 - 26], a plasmid which contains two EcoR I sites separated by 51 bp was constructed by duplication of a 192 bp fragment containing the EcoR I site of pBR322 (Fig. 2). As a result, the DNA sequences surrounding the two recognition sites are identical. DNA fragments of 1027 and 388 bp were excised from pBR322(RI)$_2$ to facilitate analysis of reaction products on polyacrylamide gels.

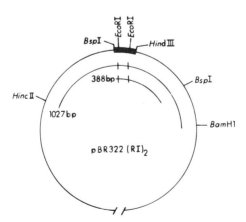

Fig. 2. Structure of pBR322(RI)$_2$. The two EcoR I sites are separated by 51 bp. Fragments of 1027 bp and 388 bp were isolated from this plasmid for kinetic studies [27].

Initially, a 1027 bp DNA substrate derived from pBR322(RI)$_2$ with the two EcoR I sites near the center of the molecule was chosen to minimize any possible end effects. The reaction of the 1027 bp DNA substrate with EcoR I endonuclease illustrates that the endonuclease cleaves this tandem site DNA in a processive manner (Fig. 3). Processive cleavage of the 1027 bp DNA decreases with increasing ionic strength (Fig. 3 and Table III). The equal production of 454 and 522 bp as well as 505 and 573 bp indicates the two EcoR I sites are cleaved with identical efficiency at all ionic strengths. Clearly, the two EcoR I sites are kinetically equivalent.

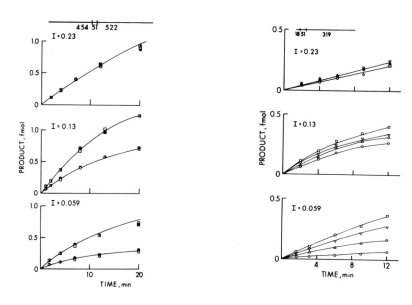

Fig. 3 (left). EcoR I endonuclease cleavage of a tandem EcoR I site DNA. Hydrolysis of a 1027 bp linear DNA fragment isolated from pBR322(RI)$_2$ was analyzed at three different ionic strengths at 37°C. Buffer conditions and fraction of processive reactions are summarized in Table III. ■, 454 bp; □, 522 bp; ●, 505 bp; ○, 573 bp products.

Fig. 4 (right). EcoR I endonuclease cleavage of 388 bp DNA. A 388 bp DNA fragment isolated from pBR322(RI)$_2$ with the two EcoR I sites asymmetrically positioned was cleaved with EcoR I endonuclease at 37°C (summarized in Table III). □, 319 bp; ∇, 18 bp; ○, 370 bp; ∆, 69 bp products.

Since DNA length has an effect on kinetic parameters [21, 22], hydrolysis of a 388 bp linear DNA fragment isolated from pBR322(RI)$_2$ was also studied. The fraction of processive cleavages for the 1027 bp and 388 bp linear DNA substrates was similar (Table III). However, when the two

EcoR I sites are located near a terminus as is the case for the 388 bp linear DNA, cleavage of the two sites differed (Fig. 4). The relative excess of 319 bp over 18 bp indicates cleavage occurs preferentially at the more centrally located EcoR I site.

Preferential cleavage at one of the EcoR I sites in the tandem site 388 bp DNA substrate could occur for a variety of reasons. Affinity of EcoR I endonuclease may be reduced for an EcoR I site located near a terminus. However, analysis of EcoR I endonuclease binding to the 388 bp tandem site DNA fragment indicated that EcoR I endonuclease binds noncooperatively to each recognition site with the same affinity [27]. In addition, the degree of processive hydrolysis of 388 and 1027 bp DNA substrates is quite similar (Table III), proving that preference for the interior EcoR I site is not due to decreased sensitivity to cleavage at the terminal site. A third possible cause for preferential hydrolysis at the more centrally located recognition site may be due to a positional advantage. If EcoR I endonuclease locates its recognition site by a facilitated diffusion mechanism, terminal placement of the two EcoR I sites produces different target sizes on each side of a recognition site. A positionally correlated transfer mechanism would favor the more centrally located site due to the larger number of nonspecific DNA binding sites on the 319 bp arm.

TABLE III. Processivity of EcoR I endonuclease[a]

[NaCl], M	Ionic strength	Fraction Processive[b]		
		1027 bp linear	388 bp linear	388 bp circle
0.025	0.059	0.38	0.46	0.77
0.10	0.13	0.18	0.16	0.34
0.20	0.23	0.00	0.03	0.03

[a] Reactions were performed in 0.02 M TrisHCl (pH 7.6), 5 mM $MgCl_2$, 0.2 mM EDTA, 0.05 mg/ml bovine serum albumin, and NaCl as indicated at 37°C.

[b] The fraction of processive cleavages are calculated from stoichiometries of singly and doubly cleaved molecules for the first 15% of the reaction [27].

Processive cleavage of the two linear DNA substrates containing tandem EcoR I sites decreases with increasing ionic strength as would be expected for a facilitated diffusion mechanism. However, only about 40% of the linear DNA substrates are cleaved in a processive manner at low ionic strength even though the DNA scan distance is about a kilobase at an ionic strength of 0.079 in the absence of Mg^{2+} [21, 22]. Topological arguments suggest that linear DNA substrates may limit processive behavior to 50% if two conditions hold: (i) there is an equal probability that the endonuclease diffuses to either side of a cleaved EcoR I site and (ii) dissociation of the cleaved EcoR I site is more rapid than translocation of the endonuclease 51 bp to the adjacent EcoR I site. In such a situation, cleavage and dissociation of the first site will leave one-half of the endonuclease molecules associated with a DNA fragment lacking the second EcoR I site, limiting processivity to a maximum of 50%.

Such topological barriers to processivity should be eliminated in circular DNA substrates since cleavage of the first site produces only a single DNA fragment. Indeed, at low and moderate ionic strength, the fraction of circular DNA cleaved in a processive reaction was approximately twice that observed for the linear DNA substrates (Table III). Processive cleavage for both linear substrates was similar implying circularization is responsible for the increase in processivity and not simply a DNA chain length effect. The two-fold increase in processivity for the circular DNA at low and moderate ionic strengths demonstrates a similar efficiency of translocation over either 51 or 337 bp to the second EcoR I site. Such a high efficiency of second EcoR I site cleavage would arise only if a positionally correlated transfer mechanism was operable.

The distance over which EcoR I endonuclease acts processively is sensitive to ionic strength as would be expected for a facilitated diffusion mechanism [28]. At low ionic strength, EcoR I endonuclease can be transferred with high efficiency between sites 300 bp apart. No processivity is observed at high ionic strength. This may indicate the distance over which the endonuclease acts processively is less than the 51 bp separating the two sites or that positionally uncorrelated transfer becomes significant at high ionic strength. Alternatively, the efficiency of recognition of an EcoR I site could be reduced at high ionic strength, ie., the endonuclease may pass over an EcoR I sequence without cleaving the site.

The dependence of cleavage kinetics on DNA chain length and demonstration of processive action indicate that EcoR I endonuclease utilizes a positionally correlated transfer mechanism during specific association and dissociation reactions under catalytic conditions. The search distance is very sensitive to the ionic strength and Mg^{2+} concentration [29]. As a result, previous reports that EcoR I endonuclease is not processive should be considered in terms of solution conditions [30, 31]. Indeed, the search distance decreases from greater than 1000 bp in 1 mM $MgCl_2$ to less than 300 bp at 10 mM $MgCl_2$ even if ionic strength is held constant [22]. Therefore, the inability to observe processivity in 20 mM $MgCl_2$ is not surprising [31]. The sensitivity of facilitated diffusion to ionic strength and divalent cation concentrations raises the question whether EcoR I endonuclease would exploit a sliding mechanism in vivo. Although the in vivo ionic strength and Mg^{2+} concentration are not precisely known [32, 33], experiments with 5 mM $MgCl_2$ at moderate ionic strength indicate that facilitated diffusion may be operative over distances of tens to hundreds of base pairs.

DNA binding proteins (HU proteins) isolated from E. coli inhibit EcoR I endonuclease cleavage at saturation (weight ratio of 2 HU/DNA) in vitro [34]. However, normal in vivo concentrations of HU proteins are only half that necessary to saturate the E. coli chromosome [35]. Nonsaturating HU serves to enhance the rate of cleavage by EcoR I endonuclease, though the cleavage rate enhancement due to DNA chain length is reduced [22]. If one assumes that the biological role of EcoR I endonuclease is to restrict foreign DNA, the in vivo significance of these observations is unclear since viral DNA entering a cell is not complexed with basic proteins [36].

IV. MECHANISM OF CLEAVAGE OF AN EcoR I SITE

Although EcoR I endonuclease specifically recognizes and cleaves at the hexanucleotide sequence (GAATTC) under normal conditions, neighboring DNA sequences have an effect on the kinetics of cleavage. Susceptibility of an EcoR I site to cleavage [18, 23, 24, 26] and whether dissociation occurs after single strand scission [25, 37, 38] are determined by external DNA sequences. Furthermore, alternative sites to the EcoR I site (EcoR I* sites) may be cleaved under certain conditions [39].

Previous kinetic studies of Modrich and Zabel suggested that the rate determining step in EcoR I endonuclease catalysis occurs after DNA cleavage [10]. Subsequent analysis of rates of cleavage and dissociation with respect to ionic strength confirmed that the rate determining step is product release from the endonuclease [15, 40]. Recently we have measured the rate of EcoR I endonuclease cleavage of pBR322 over a range of ionic strengths [B. Terry and P. Modrich, unpublished]. A representative time course for pBR322 hydrolysis by EcoR I endonuclease is given in Fig. 5. Cleavage of the first strand of pBR322 occurs at the rate of 50 min^{-1} (measured by loss of covalently closed circular DNA) and is independent of ionic strength over the range I = 0.059 - 0.23 at 37°C. Second strand cleavage occurs at 27 min^{-1} for I = 0.059 - 0.17 whereas the catalytic turnover varies from 0.72 - 4.6 min^{-1} over this range [B. Terry and P. Modrich, unpublished, 40]. Clearly, the rate determining step occurs after second strand cleavage at low and moderate ionic strengths. Even at high ionic strength where the dissociation rate increases dramatically [21], the hydrolysis steps remain faster than endonuclease turnover [W. Jack, B. Terry, and P. Modrich unpublished].

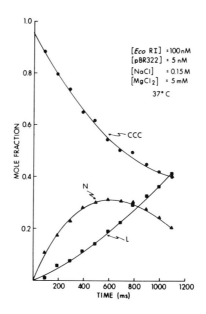

Fig. 5. *Eco*R I endonuclease cleavage of pBR322. *Eco*R I endonuclease (100 nM) was preincubated with 5 nM pBR322 at 37°C in 0.02 M TrisHCl (pH 7.6), 0.15 M NaCl, 0.5 mM EDTA, and 0.1 mg/ml bovine serum albumin. The reaction was initiated by addition of an equal volume of 0.02 M TrisHCl (pH 7.6), 0.15 M NaCl, 10 mM $MgCl_2$ utilizing a rapid mix chemical quench instrument. The reaction is quenched by the addition of two volumes 0.05 M EDTA/1% SDS. Samples were collected and analyzed by agarose electrophoresis. CCC, covalently closed circular DNA; N, nicked DNA; and L, linear DNA.

If the rate determining step occurs after DNA hydrolysis and involves nonspecific DNA sequences, small oligomers should have catalytic turnovers approaching the chemical cleavage steps. Recently, cleavage of an octamer d(pGGAATTCC) was determined to be 18 mol phosphodiester bonds hydrolyzed min^{-1} mol^{-1} endonuclease dimer (I = 0.15 at 20°C) [41]. Correcting for the lower temperature used in the oligomer studies, the turnover rate for first strand cleavage extrapolated to 37°C would be about 60 min^{-1}.

The magnitude of first and second DNA strand cleavage rates (k_2 and k_3 in Fig. 6 below) indicates that the chemical hydrolysis steps are independent of ionic strength over the range 0.059 - 0.23. In contrast, the catalytic turnover increases from 0.7 min^{-1} to 13 min^{-1} over the same ionic strength range. The increase in k_{cat} with ionic strength is consistent with a decrease in nonspecific affinity at higher ionic strength as would be expected for a facilitated diffusion mechanism. In addition, the intrinsic rate of chemical steps are similar for pBR322 (4363 bp) and a 8 bp oligonucleotide [41] implying that DNA chain length effects are manifested during association and dissociation and not during the cleavage steps. Comparison of the rates of DNA cleavage (k_2 and k_3) with rates of catalytic turnover indicates that EcoR I endonuclease spends most of the time during each catalytic cycle bound to nonspecific DNA sequences. This implies that the susceptibility of a site to cleavage is not dependent on the residence time of the endonuclease at a given DNA sequence. In addition, this suggests that the cleavage function of the enzyme is not active in a nonspecific complex.

EcoR I endonuclease binds to nonspecific DNA sequences albeit with much lower affinity than to the EcoR I site (Table II). The high degree of processivity observed at low and moderate ionic strength indicates that post-cleavage association with external sequences is a major feature of the EcoR I endonuclease pathway. To accommodate this association and also association leading up to specific complex formation, Fig. 6 presents an updated version of the proposed EcoR I endonuclease mechanism [25, 27]. Endonuclease initially associates with nonspecific DNA sequences and slides along the DNA until either the EcoR I site is located or the enzyme dissociates. Dissociation of the nicked reaction intermediate (E·1) may occur with certain DNA substrates, at low temperature, or in the presence of DNA intercalators [25, 38, 42]. In the presence of Mg^{2+}, the EcoR I endonuclease dimer cleaves both strands of pBR322 efficiently during a single binding event [40]. Subsequent to EcoR I site cleavage the endonuclease remains associated with nonspecific DNA sequences. If a second EcoR I site is nearby, the second site may be cleaved in a processive manner.

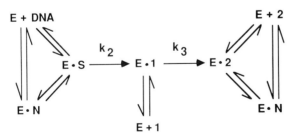

Fig. 6. EcoR I endonuclease reaction mechanism. E·N and E·S refer to nonspecific and specific endonuclease DNA complexes, respectively. The reaction intermediate 1 is hydrolyzed in only one DNA strand, while 2 represents the DNA product cleaved in both strands.

What is the nature of the rate determining step? We believe the rate determining step is dissociation of endonuclease from nonspecific DNA sequences [15, 27, 30]. Processivity experiments indicate that EcoR I endonuclease does not dissociate directly from the cleaved EcoR I site at low and moderate ionic strengths. Comparison of endonuclease cleavage rates with rates of dissociation in the absence of Mg^{2+} at an ionic strength of 0.079 are consistent with this assignment [40] as is the finding that k_{cat} for cleavage of octamer substrates [41] is similar to the rate of the chemical steps determined with pBR322.

The above conclusions differ from those of Halford and colleagues who have suggested that a rate determining conformational change is associated with specific site recognition under catalytic conditions on a pMB9 substrate [17, 38]. Due to substrate differences, a direct comparision is difficult. It is pertinent, however, to note the fluorescence assay employed in the pMB9 studies is independent of steps occurring subsequent to topological release of one DNA terminus generated as a consequence of first strand scission by the endonuclease.

V. REFERENCES

1. Modrich, P. (1982) CRC Crit. Rev. Biochem. 13, 287-323.
2. Hedgpeth, J., Goodman, H.M., and Boyer, H.W. (1972) Proc. Natl. Acad. Sci. U.S.A. 80, 31-35.
3. Dugaiczyk, A., Hedgpeth, J., Boyer, H. W., and Goodman, H. W. (1974) Biochemistry 13, 503-512.
4. Greene, P. J., Poonian, M. S., Nussbaum, A. L., Tobias, L., Garfin, D. E., Boyer, H. W., and Goodman, H. M. (1975) J. Mol. Biol. 99, 237-261.
5. Rubin, R. A. and Modrich, P. (1977) J. Biol. Chem. 252, 7265-7272.
6. Goppelt, M., Pingoud, A., Maass, G., Mayer, H., Koster, H., and Frank, R. (1980) Eur. J. Biochem. 104, 101-107.
7. Lu, A.-L., Jack, W. E., and Modrich, P. (1981) J. Biol. Chem. 256, 13200-13206.
8. Frederick, C. A., Grable, J., Melia, M., Samudzi, C., Jen-Jacobson, L., Wang, B.-C., Greene, P., Boyer, H. W., and Rosenberg, J. M. (1984) Nature 309, 327-330.
9. Halford, S. E. and Johnson, N. P. (1980) Biochem. J. 191, 593-604.
10. Modrich, P. and Zabel, D. (1976) J. Biol. Chem. 251, 5866-5874.
11. Modrich, P. (1979) Q. Rev. Biophys. 12, 315-369.
12. Jack, W. E., Rubin, R. A., Newman, A., and Modrich, P. (1981) in Gene Amplification and Analysis, (Chirikjian, J. G., ed.) Vol. 1, pp. 165-179, Elsevier/North-Holland, New York.

13. Rosenberg, J. M., Boyer, H. W., and Greene, P. J. (1981) in Gene Amplification and Analysis (Chirikjian, J. G., ed.) Vol. 1, pp. 131-164, Elsevier/North-Holland, New York.
14. Terry, B. J., Jack, W. E., Rubin, R. A., and Modrich, P. (1983) J. Biol. Chem. 258, 9820-9825.
15. Jen-Jacobson, L., Kurpiewski, M., Lesser, D., Grable, J., Boyer, H., Rosenberg, J. M., and Greene, P. J. (1983) J. Biol. Chem. 258, 14638-14646.
16. McGhee, J. D. and von Hippel, P. H. (1974) J. Mol. Biol. 86, 469-489.
17. Halford, S. E. and Johnson, N. P. (1983) Biochem. J. 211, 405-415.
18. Halford, S. E., Johnson, N. P., and Grinsted, J. (1980) Biochem. J. 191, 581-592.
19. Richter, P. H. and Eigen, M. (1974) Biophys. Chem. 2, 255-263.
20. Berg, O. G., Winter, R. B., and von Hippel, P. H. (1981) Biochemistry 20, 6929-6948.
21. Jack, W. E., Terry, B. J., and Modrich, P. (1982) Proc. Natl. Acad. Sci. USA 79, 4010-4014.
22. Ehbrecht, H., Pingoud, A., Urbanke, C., Maass, G., and Gualerzi, C. (1985) J. Biol. Chem. 260, 6160-6166.
23. Thomas, M. and Davis, R. W. (1975) J. Mol. Biol. 91, 315-328.
24. Forsblom, S., Rigler, R., Ehrenberg, M., Petterson, U., and Philipson, L. (1976) Nucleic Acids Res. 3, 3255-3269.
25. Rubin, R. A. and Modrich, P. (1978) Nucleic Acids Res. 5, 2991-2997.
26. Alves, J., Pingoud, A., Haupt, W., Langowski, J., Peters, F., Maass, G., and Wolff, C. (1984) Eur. J. Biochem. 140, 83-92.
27. Terry, B. J., Jack, W. E., and Modrich, P. (1985) J. Biol. Chem. 260, 13130-13137.
28. Winter, R. B. and von Hippel, P. H. (1981) Biochemistry 20, 6948-6960.
29. Lohman, T.M. (1985) CRC Crit. Rev. Biochem. 191, 191-245.
30. Rubin, R. A. and Modrich, P. (1980) in Methods in Enzymology (Grossman, L. and Moldave, K., eds.) Vol. 65, pp. 96-104, Academic Press, New York.
31. Langowski, J., Alves, J., Pingoud, A., and Maass, G. (1983) Nucleic Acids Res. 11, 501-513.
32. Kao-Huang, Y., Rezvin, A., Butler, A.P., O'Conner, P., Noble, D.W., and von Hippel, P.H. (1977) Proc. Natl. Acad. Sci. U.S.A. 74, 4228-4232.
33. Lovgren, T.N.E., Petersson, A., and Loftfield, R.B. (1978) J. Biol. Chem. 253, 6702-6710.
34. Pingoud, A., Urbanke, C., Alves, J., Ehbrecht, H., Zabeau, M., and Gualerzi, C. (1984) Biochemistry 23, 5697-5703.

35. Rouviere-Yaniv, J. and Gros, F. (1975) Proc. Natl. Acad. Sci. U.S.A. 72, 3428-3432.
36. Hershey, A.D. and Chase, M. (1952) J. Gen. Physiol. 36, 39-52.
37. Halford, S. E., Johnson, N. P., and Grinsted, J. (1979) Biochem. J. 179, 353-365.
38. Halford, S. E. (1983) Trends in Biochem. Sci. 8, 455-460.
39. Polisky, B., Greene, P., Garfin, D. E., McCarthy, B. J., Goodman, H. M., and Boyer, H. W. (1975) Proc. Natl. Acad. Sci. USA 72, 3310-3314.
40. Jack, W. E. (1983) Dissertation (Duke University, Durham, NC).
41. Brennan, C.A., Van Cleve, M.D., and Gumport, R.I. (1986) J. Biol. Chem. 261, 7270-7278.
42. Halford, S. E. and Johnson, N. P. (1981) Biochem. J. 199, 767-777.

4

Structure and Function of the
EcoR I Restriction Endonuclease

John M. Rosenberg[1], Judith A. McClarin[1], Christin A. Frederick[1,2]
John Grable[1], Herbert W. Boyer[3], and Patricia J. Greene[3]

[1]Department of Biological Sciences
University of Pittsburgh
Pittsburgh, PA 15260

[2]Department of Biology
Massachusetts Institute of Technology
77 Massachusetts Avenue
Cambridge, MA 02139

[3]Department of Biochemistry and Biophysics
University of California, SF
San Francisco, CA 94143

INTRODUCTION

The highly specific recognition of the double-stranded sequence d(GAATTC) by EcoR I endonuclease offers compelling advantages as a system for investigating sequence specific recognition of DNA. It is a small protein (31,065 daltons, 276 amino acids) of known sequence [1, 2] which forms highly stable catalytically active dimers in solution [3, 4]. The enzyme hydrolyzes the phosphodiester bond between the guanylic and adenylic acid residues resulting in a 5'-phosphate. The reaction proceeds with inversion of configuration at the reactive phosphorus [5], implying that there is an odd number of chemical events during the hydrolysis. The simplest interpretation of this observation is that the enzyme does not form a covalent intermediate with the DNA. Although EcoR I endonuclease requires Mg^{2+} for phosphodiester bond hydrolysis, it binds specifically to its cognate hexanucleotide in the absence of Mg^{2+} with a dissociation constant on the order of 10^{-11} M^{-1} [6, 7, 8, 9].

The enzyme also binds DNA in a nonspecific manner ie., at sites other than GAATTC; this does not result in hydrolysis of the DNA [6, 10, 11]. It has been postulated that the nonspecific complex enhances the rate of formation of formation of the specific complex by facilitated diffusion along the DNA [9, 12, 13, 14].

Both the EcoR I endonuclease and the EcoR I methylase recognize the same hexanucleotide; however, the latter methylates the central adenine residues of both strands at the exocyclic N-6 amino group. When either one or both groups is methylated the endonuclease no longer cleaves the DNA. Thus, EcoR I endonuclease not only discriminates between its hexanucleotide and all other hexanucleotides, it also discriminates between different methylation states of the same hexanucleotide.

Cocrystals of DNA and protein that diffract to high resolution are required for a full understanding of sequence specificity in the EcoR I system. We have obtained cocrystals and determined their structure [15, 16, 17]. Here we report he structure of the recognition complex including interactions involved in sequence specificity.

STRUCTURE OF THE DNA-ECOR I ENDONUCLEASE COMPLEX

General features of the complex

Both subunits of the enzyme form a globular structure with the DNA embedded in one side (Fig. 1). The complex as a whole is approximately 50 Å across. The major groove of the DNA is in intimate contact with the protein while the minor groove is clearly exposed to solvent. The complex has twofold symmetry, as expected from the symmetry of the recognition sequence. The protein has two projecting features, termed arms, that wrap around the DNA.

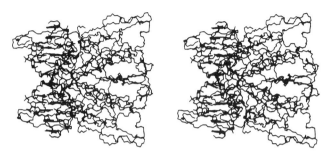

Figure 1. A stereo drawing of the main chain atoms of the protein and the non-hydrogen atoms of the DNA, which is towards the left of the complex. The twofold axis of rotationaly symmetry is horizontal and in the plane of the figure.

Structure of the DNA Within the Complex

The DNA retains most of the structural features of the well-known double helix. In particular, Watson-Crick base pairing is maintained throughout the 12 paired bases. However, the DNA is kinked in the recognition complex, by which we mean that it departs significantly from the B conformation according to certain criteria [17]. Each kink distorts approximately two base pairs and the centers of the kinks are separated by three base pairs (Figs. 2 and 3). These kinks appear to be stabilized by the binding of the protein.

The most striking departure from B-DNA is centered on the crystallographic and molecular twofold axis, between adenine-6 and thymine-7 (Fig. 2). We refer to this feature as the "type I neokink". It represents a net rotation of the upper half of both strands of the DNA relative to the entire lower half of the double helix so as to unwind the DNA. The unwinding can be seen in the relative positions of phosphorus atoms 6 and 7, which show very little relative rotation about the average helix axis (they are at the center of Fig. 3). The unwinding is approximately 25° and would propagate through the DNA as a long-range effect on the net winding of the double helix. Kim and co-workers have measured the unwinding of DNA in solution when EcoR I endonuclease binds DNA in the absence of Mg^{2+}; they obtained an identical value [18].

```
0   1   2   3   4   5   6   7   8   9  10 11 12
T  pC  pG  pC  pG  pA  pA  pT  pT  pC  pG  pC pG
    |           |           |
  Type II     Type I     Type II
  neo-kink   neo-kink   neo-kink
```

Figure 2. The sequence of the tridecameric oligonucleotide used to make the DNA-protein complex. Also shown is the location of the kinks and the base numbering scheme, which was chosen to be consistent with the numbering system used by Dickerson and co-workers for the dodecamer [35, 36, 37]; thus a given residue, eg. guanine 2 refers to the same residue in both the dodecamer and the EcoR I complex.

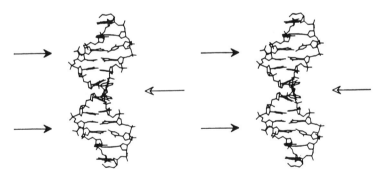

Figure 3. A stereo figure of the DNA indicating the type I and type II neokinks. The single arrow on the right (with an open head) points to the center of the type I neokink. The twofold symmetry axis passes through both the arrow and the center of the type I neokink; hence the type I neokink has this symmetry. The two arrows on the left (with filled heads) point to the centers of the type II neokinks; they are identical because of the twofold symmetry.

The principal effect of the unwinding is that the major groove becomes wider. The phosphate-phosphate distances across the major groove are increased by approximately 3.5 Å. Interestingly, the base pairs do not significantly increase their interplanar separation although the base-base stacking contacts are clearly changed (Fig. 3). Thus, the type I neokink represents an effective mechanism for increasing the separation of the backbones of DNA strands without increasing the separation of the bases. The difference arises because a helix is a screw. Breaking the screw symmetry at one point of a helix and twisting one part with respect the other will alter the separation between the "threads" across the break. The increased backbone separation is essential because otherwise the recognition α-helices would not fit in between and therefore could not approach closely enough to interact with the bases (see below). This consequence of the type I neokink suggests that it may be a general mechanism for facilitating access by proteins to the major groove of DNA. If so, similar DNA structures should be seen in some other recognition complexes.

There are also significant displacements of the A·T base pairs on either side of the kink center. These base displacements are critical to the recognition mechanism because they align adjacent adenine residues (five and six) within the recognition site (Fig. 2). These two purines are both involved in "bridging" interactions with amino acid side chains. These recognition interactions could not occur without the realignment because the N-6 moieties bridged by Glu^{144} and the N-7 moieties bridged by Arg^{145} would be too far apart if the DNA were in the B conformation.

Both the base pair realignment and the increased backbone separation are manifestations of a localized reduction in the twist of DNA; hence one could probably not exist without the other. However, the unwinding between adjacent phosphates appears to be localized at residues that are different from those where unwinding appears between adjacent base pairs. The phosphate unwinding is concentrated at the middle of the DNA (between phosphates 5' to residues 6 and 7), whereas the base unwinding is displaced toward the adenines (residues 5 and 6), thus producing the realignment discussed above. Thymines 7 and 8, which are paired to displaced adenines, are also displaced.

The realignment of the pairs reveals another aspect of the type I neokink that may be of general significance: namely, that it creates sites for multiple hydrogen bonds which are absent in B-DNA. Indeed, the idea that *Eco*R I endonuclease creates some of the detailed features on the surface of the DNA, which it then recognizes, is provocative and unexpected.

We determined the helical properties for the segments of DNA between the neokinks and between the type II neokink and the end of the DNA [19], using the coordinates we fit to the ISIR electron density map. We obtained results similar to the corresponding determinations based on the preliminary DNA model [16]: The bend angle of the type II neokink is between 20° and 40°. However, the interpretation of this result is clouded because these calculations include nucleotides that are not in an exact helical conformation. Highly refined coordinates (which are not yet available) are required to properly choose which to include in the calculations. Consequently, values for the bend angle of the type II neokink should be considered provisional. Unwinding can be more readily assessed by examining phosphorus positions in projection down the average helix axis, and the type II neokink does not introduce a major change in the net winding of the DNA.

Structural Organization of the Protein

Each EcoR I endonuclease subunit is organized into a single domain consisting of five-stranded β sheet surrounded on both sides by α helices (see Fig. 4). The domain is therefore of the well-known α/B architecture [20]. Four of the five strands in the β sheet are parallel; however the location of the single antiparallel strand makes it possible to divide the sheet conceptually into parallel and antiparallel three-stranded motifs. The parallel motif ($\beta 3$, $\beta 4$, and $\beta 5$ of Fig. 4) is the foundation of the direct contacts between the protein and DNA bases as well as subunit-subunit interaction, and the antiparallel motif ($\beta 1$, $\beta 2$, and $\beta 3$ of Fig. 4) is the foundation for the site of DNA strand scission. The parallel motif is very similar to one-half of the well-known nucleotide binding domain [21].

All of the major α helices in the protein are aligned so that their amino-terminal ends are pointing in the general direction of the DNA. This orients the α helix dipoles so that they interact favorably with the electrostatic field generated by the negatively charged phosphates on the DNA backbone, thereby contributing to the net stability of the complex. (Because of the alignment of the peptide bonds, an α helix has a net dipole moment which can be approximated by placing one-half of a positive virtual charge at the amino terminus of the helix and one-half of a negative virtual charge at the carboxyl terminus [22]). The inner and outer α helices from each subunit are oriented so that their amino-terminal ends project into the major groove of the DNA. The amino acid side chains that interact with the DNA bases are located at the ends of these helices or in residues that immediately precede the helix.

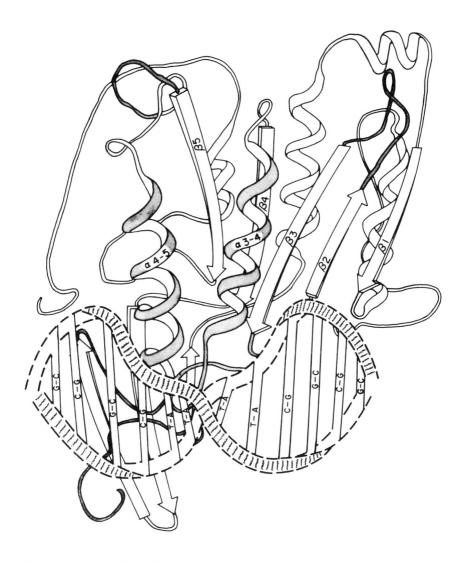

Figure 4. Schematic backbone drawing of one subunit of (dimeric) EcoR I endonuclease and both strands of the DNA in the complex. The arrows represent β strands, the coils represent α helices and the ribbons represent the DNA backbone. The helices in the foreground of the diagram are the inner and outer recognition helices. They connect the third β strand to the fourth and the fourth β strand to the fifth. The two helices also form the central interface with the other subunit. The amino-terminus of the polypeptide chain is in the arm near the DNA.

Subunit-subunit interactions are primarily mediated by amino acid residues located in the parallel motif. The subunit-subunit interface can be subdivided into two general regions: A central portion, which is inaccessible to solvent, and a surface portion, which is solvent accessible. The central portion of the extensive interface includes interactions between residues within the two crossover a helices, that is, the inner and outer a helices. The NH_2-terminus of β-strand 5 is also part of the central interface. These interfacial residues have hydrophobic and neutral polar side chains. The surface portion of the interface includes many salt links between subunits situated around the exterior edge of the interface. These are formed by charged residues in the turn preceding β-strand 5, residues in the carboxy-terminal surface loop, and two residues from the surface of the antiparallel motif (all other residues in the interface are in the parallel motif). Charged residues at the subunit-subunit interface are also involved in the DNA-protein interface.

EcoR I endonuclease has arms that wrap around the DNA (Fig. 1). Each arm is an extension of the α/β domain which wraps around the DNA partially encircling it, thereby clamping it into place on the surface of the enzyme. Because of the twofold symmetry of the complex there are two arms, which interact with the DNA directly across the double-stranded helix from the scissile bonds. They contact the DNA at the type II neokinks and may be causative elements in the formation of these DNA structures. Each arm is composed of the amino terminus of the protein and a β hairpin sequentially located between the fourth and fifth strands of the large β sheet (residues 176 through 192)*. Part of the amino-terminal portion of the polypeptide chain (residues 17 through 20 adds a third β strand to the β hairpin, thereby forming a three-stranded antiparallel β sheet, which is the structural foundation of the arm. Thus, there are two β sheets in each EcoR I endonuclease subunit: The large five-stranded sheet described above and the smaller three-stranded sheet described here.

The first 14 amino acid residues of the polypeptide chain form an irregular structure which is sandwiched in between the smaller β sheet and the DNA. The sandwiched region of the arm mediates several nonspecific DNA-protein contacts. Additional DNA backbone contacts are located in the short segment of polypeptide chain that connects the β hairpin with the outer α helix, which follows it in the primary sequence.

---

*The exclusion of the beta-hairpin from the primary topology of the domain derives from the convention that protuberances of this type be excluded from the assignment of the basic topological elements of a domain [20] since they could represent the evolutionary consequences of an insertion of DNA into an ancestral gene at a point which coded for a loop at the protein surface.

Catalytic Clefts in the Enzyme

The two DNA backbone segments that face toward the major portion of the endonuclease are buried in clefts in the protein. These segments include the scissile bonds. Both DNA backbone segments and the corresponding clefts are identical because of the twofold symmetry of the complex. The carboxyl edge of the antiparallel segment of the β sheet forms the base of the cleft which binds phosphates three, four, and five (Fig 2). (The scissile bond is at the fifth phosphate). One side of the catalytic cleft is formed by the loops which interconnect the β strands in the antiparallel motif and that connect β strand 3 to the inner α helix. The scissile bond is facing this side of the cleft. The other side of the cleft is formed by the inner and outer a helices from the other subunit. The cleft surface contains many basic amino acid residues which interact electrostatically with the DNA phosphates, contributing to the binding energy.

It has been known for some time that Mg^{2+} can be added to preformed EcoR I endonuclease-DNA complexes in solution, which are then activated for cleavage [3], that is, the order of addition can be first DNA and then Mg^{2+}. We therefore diffused Mg^{2+} into the cocrystals and found that the hydrolytic reaction was carried out in the crystalline state [23]. This demonstrates the catalytic competence of the crystalline DNA-protein complex. The Mg^{2+}-treated crystals survive the structural transitions and they still diffract X-rays. The structure of the enzyme-product complex is not yet known.

The active site for DNA strand cleavage is not fully assembled in our structure. There is a solvent channel, with DNA backbone on one side and protein on the other, ending at the scissile bond. It is through this solvent channel that magnesium probably enters the active site. We presume that the structure in this region rearranges after magnesium is bound, forming a functional active site. In other words, in the absence of Mg^{2+}, the complex is analogous to an inactive zymogen that is activated by a structural isomerization triggered by the cation. This temporal order is probably important in the function of the endonuclease (see below).

DNA Backbone-Protein Interactions

There appear to be interactions between the protein and the backbone of the DNA from residues two through the nine (Fig. 2). Phosphate moieties from residues 3, 4, and 7 are buried in the protein and are inaccessible to solvent. Phosphate and deoxyribose residues 3, 4, and 5 are on each strand line the sides of the recognition hexanucleotide major groove, which is expanded by the type I neokink. These phosphates are bound within the catalytic clefts in the protein. Electrostatic interactions are also formed between the arms of the protein and phosphates from residues 8 and 9.

The two symmetrically related clefts, one in each subunit, are approximately 3.5 Å farther apart than the normal separation between the DNA backbones across the major groove of B-DNA. The increased separation, coupled with the basic residues within the clefts, probably produces an electrostatic field, which would tend to drive the DNA backbones apart. This could be a major factor promoting the formation of the type I neokink.

Phosphates 3 and 4, which flank the type II neokink are not only buried in the cleft but interact with several basic amino acid residues. The strong interaction could be associated with a requirement to precisely position the scissile bond in the active site of the enzyme.

Ethylation interference experiments showed the largest effects at phosphate moieties of residues 3, 4, and 7 [24], that match the phosphates that are buried in the protein and protected from solvent. The next largest ethylation effect is observed for the reactive phosphate at the fifth position. Small effects are noted for the sixth phosphate, which is probably forming interactions to the protein even though it is partially exposed to the solvent. (We suspect that a stronger ethylation interference would have been observed at lower protein concentrations where the equilibrium is sensitive to smaller reductions in the protein-DNA association constant).

The oligonucleotide used in our cocrystal is long enough to include all of the major contacts that form between long DNA substrates and the endonuclease. The association constant measured for the dodecamer CGCGAATTCGCG is within experimental error of that measured for plasmid DNA [2, 4, 9, 25]. The unusually large Michaelis constant for an octanucleotide substrate as compared with dodecameric or larger substrates [4, 26] suggests that interactions between the enzyme and the flanking regions of the DNA backbone make significant contributions to the net stability of the complex.

The loop connecting β strand 3 with the inner α helix (residues 131-143) appears to have a pivotal role in facilitating structural communication between regions of the complex. Part of this loop is involved in the formation of the cleavage site, as indicated above. However, its three-dimensional neighbors include many vital components of the complex. This loop is simultaneously adjacent to the arm, close to the amino acid residues involved in the direct recognition; the region around residue 140 is also packed against phosphate 7, which is the center of the type I neokink. The structure of these residues could be directly influenced by (i) the formation of direct hydrogen bonds to the bases, (ii) formation of the type I neokink, (iii) the conformation of the arm, and (iv) the conformation of the cleavage site. Therefore the 131-143 loop could transmit structural information between these sites thereby serving to facilitate a temporal ordering of events within the overall catalytic cycle.

THE RECOGNITION MECHANISM

Overview

The DNA-protein interface can be viewed in two portions: An extensive interface between the protein and the backbone of the DNA (already discussed) and a protein-base interface that partially covers the major groove of the recognition hexanucleotide (GAATTC). The minor groove is open to the solvent. The protein-backbone interface spans more nucleotide residues than the protein-base interface; that is, the protein interacts with the phosphate and deoxyribose moieties from nucleotides adjacent to the canonical hexanucleotide.

Hydrogen bonds between amino acid side chains (Glu^{144}, Arg^{145}, and Arg^{200}) and the purine bases of the canonical hexanucleotide constitute the direct, sequence specific DNA-protein interactons in the complex. The bases and amino acid side chains must be precisely positioned relative to each other so that the interactions between them can generate the correct specificity: α helical motifs form critical structural elements that facilitate the establishment of that spatial juxtaposition. Different recognition α helices provide the structural foundation for the interaction with different sections of the canonical hexanucleotide GAATTC. An inner module recognizes the inner tetranucleotide AATT, and two identical symmetry-related outer modules recognize the outer G·C base pairs.

The interaction involving the outer module is relatively simple (Figs. 5 and 6): The guanidinium moiety of Arg^{200} forms two hydrogen bonds with the guanine base in an interaction designated Arg::G. One hydrogen bond is donated by Arg^{200} to the guanine N-7 atom and another is donated to the O-6 atom (see Fig. 7 for the numbering system). The Arg::G interaction was predicted by Seeman, Rosenberg, and Rich [27].

The inner module forms a more complicated set of interactions in that pairs of amino acid side chains interact with pairs of adjacent adenine residues. Each pair of adjacent adenines interacts with one amino acid from each subunit, (Fig. 5). These residues are Glu^{144} and Arg^{145} and the interaction is designated Glu-Arg::AA.

The side chain of Glu^{144} receives two hydrogen bonds from the adenine N-6 amino groups. In our current model, both hydrogen bonds are to the same carboxyl oxygen atom. (Each carboxyl oxygen atom can receive two hydrogen bonds). The second oxygen atom may be interacting with residues Arg^{200} or Arg^{203} (or both) of the outer module via water bridges. Arg^{145} donates two hydrogen bonds to the adenine N-7 atoms. Thus, recognition in the inner tetranucleotide is based on "bridging" interactions in which amino acid side chains interact with two adjacent bases.

The inner α helix is oriented so that its amino-terminal end points toward the major groove of the DNA. It makes an angle of approximately 60° with the average DNA helix axis (Fig. 6A). The polypeptide chain turns sharply at the end of the α helix so that residues at the amino-terminal end of the helix and those in the bend are in close proximity to the DNA. The amino terminus of the inner α helix is also adjacent to the molecular twofold axis and therefore it is in close proximity to the amino terminus of the symmetry related helix from the other subunit. This symmetric pair of helices together form the inner module (Fig. 6B).

Both of the outer modules consist of a single α helix, namely the outer α helix, which connects the fourth and fifth strands of the large β sheet (Fig. 6C). The inner and outer α helices are positioned somewhat differently with respect to the DNA; the helix axis of the inner α helix almost intersects the average DNA helix axis while the axis of the outer α helix passes well to the outside (Fig 6D).

The recognition α helices illustrate a principle of "positioning", that refers to all the structural factors responsible for the precise three-dimensional juxtaposition of the correct elements of the protein and the DNA. For example, Arg^{145} on the inner module recognizes features of the DNA that are different from those recognized by Arg^{200} on the outer module, because the inner and outer helices are positioned differently with respect to the DNA.

EcoR I endonuclease repeatedly utilizes a simple structural motif, the α helix, to interact with DNA. These structural motifs are part of larger topological motifs which are repeated; both are part of β-α-β units. Thus, simple repeated motifs form the structural foundation of the recognition interactions.

All four helices form a parallel helix bundle (Fig. 6D), which is stabilized by interactions between the side chains of the individual helices. This parallel α-helical bundle is significantly different from the common four-helical motif referred to as an "up and down" or antiparallel helix bundle by Richardson [20]. The antiparallel bundle has been observed in proteins such as the tobacco mosaic virus coat protein and myohemerythrin. The antiparallel architecture produces an internally favorable interaction between the electric dipoles associated with each α helix. By contrast, there is an internal energetic penalty associated with the parallel arrangement of the helices in EcoR I endonuclease. However the parallel arrangement produces an electrostatic field that facilitates the DNA-protein interaction. Another point of comparison is that the antiparallel bundle is formed by a contiguous stretch of polypeptide chain with turns connecting the helices. It is therefore a stable domain that can and does constitute the bulk of a globular protein. The parallel helix bundle must, of necessity, be part of a larger structural unit since additional elements of secondary structure are needed to connect the ends of the helices.

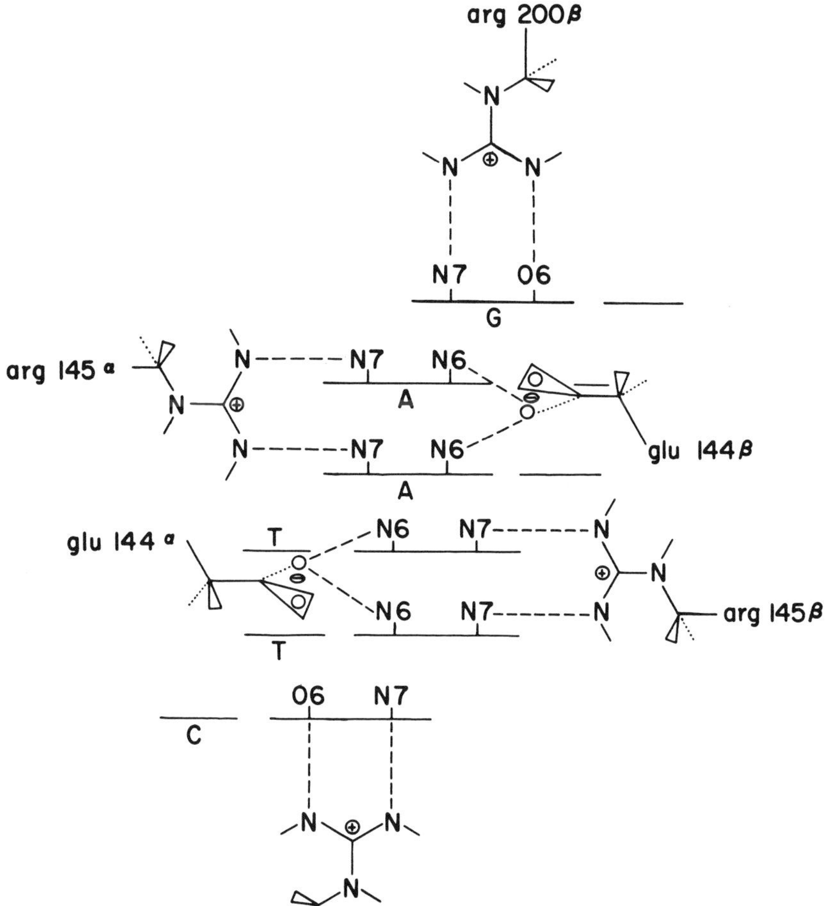

Figure 5. A schematic representation of the recognition interactions and the 12 hydrogen bonds which determine the specificity of *Eco*R I endonuclease. Here, α and β refer to the two identical subunits of the enzyme. The positions of the bases and amino acid side chains have been shifted from the current model as shown in Fig. 6 in the interests of clarity.

The inner and outer α helices determine the positions of key amino acid side chains with respect to the bases; the placement of the α helices with respect to the DNA is determined in part by side-chain interactions between the helix bundle and the DNA backbone (Fig. 6). The α helices are also packed against the β sheets of their respective subunits, thus firmly fixing the location of the entire four-helix bundle with respect to the protein as a whole. Thus, all the interactions between the protein and the DNA backbone indirectly serve to locate the helix bundle with respect to the DNA.

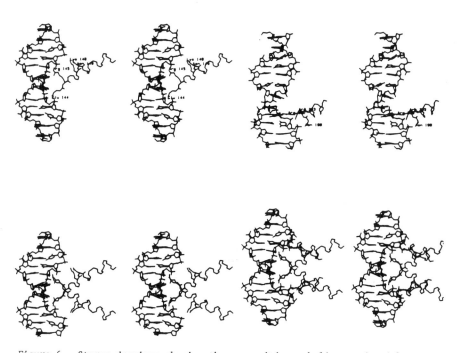

Figure 6. Stereo drawings showing the recognition α helices and modules. a). The "inner" α helix, which is part of the inner recognition module. The inner helix is also a crossover helix, connecting the third and fourth strands of the β sheet. Glu^{144} interacts with adenine residues in the lower half of the DNA and Arg^{145} interacts with adenine residues in the upper half. Lys^{148} and Asn^{149} interact with the phosphate moiety from guanine 4. b). The inner recognition module, consisting of the inner α helices from both subunits. The inner module determines the specificity in the inner tetranucleotide, AATT. c). The "outer" α helix, which is also one of two identical outer recognition modules. The outer helix connects the fourth and fifth β strands. Arg^{200} interacts with guanine. In the outer views in this figure, the twofold symmetry axis is in the plane of the drawing; however, this view has been rotated approximately 20° for clarity. Asn^{199} interacts with phosphate moiety from cytosine 3' (C3 on the opposite strand), while Arg^{203} interacts with phosphate moieties from cytosine 3' and guanine 4'. d). The four helix bundle consisting of the inner and outer α helices from both subunits.

Twelve Hydrogen Bonds Provide Sequence Specificity

There are two protein-base hydrogen bonds associated with each of the two Arg::G interactions of the two outer modules and eight hydrogen bonds associated with the inner module; four from each of the two Glu-Arg::AA interactions. This gives a total of 12 hydrogen bonds between protein and bases.

The central function of any sequence specific protein is its ability to discriminate its cognate sequence from the vast excess of noncognate DNA sequences in which it is embedded. Hence, and putative structural model of a sequence specific interaction between DNA and protein must provide a satisfactory answer to the question of what happens when noncognate bases are present in the binding site of the protein. It is therefore vital to determine wheter or not the 12 hydrogen bonds discriminate between the EcoR I site and all other possible hexanucleotides. The following discussion shows that they do because substitution of any noncognate base pair would rupture one or more hydrogen bonds. A secondary purpose of that discussion is to develop a systematic method for analyzing the sequence specificity of particular protein-DNA interactions, which is based on the ideas of Seeman, Rosenberg and Rich [27]. They showed that there were four principal interaction sites on the major groove side of a base pair and three sites on the minor groove side[5]. These specificity sites (Fig. 7) make it possible to specify a template for the DNA-protein interface that can be used to analyze the match between the protein and alternative DNA sequences.

The templates can be specified and analyzed systematically for all possible combinations of base pairs [28] (Tables 1 to 3). Consider the Arg::G interaction: Guanine has hydrogen bond acceptors in W1 and W2, which are matched by corresponding hydrogen bond donors on Arg[200]. No other base pair has hydrogen bond acceptors in both W1 and W2 (Table 1). The analysis for the Glu-Arg::AA interaction is similar. Both adenine residues have hydrogen bond acceptors on Arg[145] and Glu[144], respectively. None of the other bases match this pattern (Table 1).

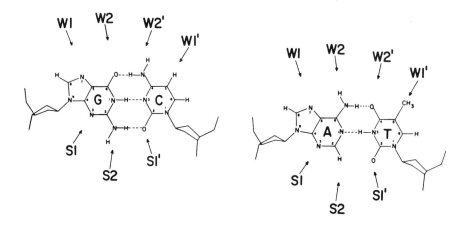
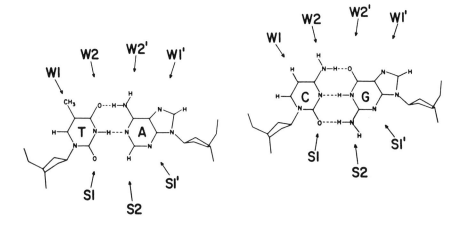

Figure 7. The four base pairs showing the numbering scheme and the specificity sites which are based on those proposed by Seeman, Rosenberg and Rich [27]. Sites W2 and W3 of Seeman, Rosenberg and Rich, which are 1 Å apart, have been merged in the current treatment and are shown here as W2. Similarly, W2' and W3' of Seeman, Rosenberg and Rich are combined into W2' here (see text).

Table 1. Base Pair Specificity Sites.

Base Pair	Contents of Site						
	W1	W2	W2'	W1'	S1	S2	S1'
A·T	A	D	A	M	A	H	A
G·C	A	A	D	H	A	D	A
C·G	H	D	A	A	A	D	A
T·A	M	A	D	A	A	H	A
^{me}A·T	A	M	A	M	A	H	A
T·^{me}A	M	A	M	A	A	H	A
G·^{me}C	A	A	D	M	A	D	A
^{me}C·G	M	D	A	A	A	D	A
A·U	A	D	A	H	A	H	A
U·A	H	A	D	A	A	H	A

The contents of the specificity sites on common base pairs are listed. Methylated bases are included as well as uracil for RNA: ^{me}A, N-6 methyl adenine; ^{me}C, 5 methyl cytosine. W1, W2 etc. are the sites shown in Fig. 7. A, a hydrogen bond acceptor is present on the indicated base pair at the indicated site; D, a hydrogen bond donor is present; M, a methyl group is present; H, a hydrogen atom (C-H) is present.

Table 2. Base Pairs Recognized by Single Interactions.

Site	Occupied by*	Symbol[c]	Base Pairs Recognized
W1	D_p	Pu	A·T, G·T, (^{me}A·T, G·^{me}C, A·U)
W1	V_o	Me	T·A, (^{me}C·G, T·^{me}A)
W1	V_i	CU	C·G, (U·A)
W2	D_p	Gt	G·C, T·A, (G·^{me}C, T·^{me}A, U·A)
W2	A_p	Ac	A·T, C·G, (^{me}C·G, A·U)
W2	V_o	MA	(^{me}A·T)
W2'	D_p	Ac'	A·T, C·G, (^{me}A·T, ^{me}C·G, A·U)
W2'	A_p	Gt'	G·C, T·A, (G·^{me}C, U·A)
W2'	V_o	MT	(T·^{me}A)
W1'	D_p	Py	T·A, C·G, (T·^{me}A, ^{me}C·G, U·A)
W1'	V_o	AM	A·T, (^{me}A·T, G·^{me}C)
W1'	V_i	GV	G·C (A·U)
S1	D_p	N	All base pairs
S2	V_i	At	A·T, T·A, (^{me}A·T, T·^{me}A, A·U, U·A)
S2	A_p	Gc	G·C, C·G, (G·^{me}C, ^{me}C·G)
S1'	D_p	N	All base pairs

As can be seen in Table 1, any given site is generally occupied by the same functional group on more than one base pair. Thus, as noted by Seeman, Rosenberg and Rich [27], a single protein-base interaction would lead to recognition of a degenerate set of base pairs. The base pairs recognized by such singular interactions are listed. *The column indicates the functional group on a hypothetical protein that interacts with the base pair in the indicated specificity site: D_p indicates that a hydrogen bond donor is present on the protein, which would be paired to an acceptor on the base; A_p indicates that a hydrogen bond acceptor is present on the protein; V_o indicates the presence of an "outer" Van der Waals contact to a methyl group on the DNA; V_i indicates the presence on an "inner" Van der Waals contact to a C-H hydrogen on a base. Here, "outer" and "inner" refer to the distance between the protein side chain and the base pair. A hydrophobic amino acid side chain in W1 would code for T or ^{me}C if it were positioned just far enough from the base pair to contact the methyl group. However, if it were positioned closer to the base pair, it would code for C by contacting C_5-H. [c]The symbol in this column refers either to the site/interaction combination or to the degenerate set of base pairs recognized. For example, Pu refers to a hydrogen bond donor on a protein in W1 which, in effect, codes for purines.

Table 3. Combinations Giving Unambigous Base Recognition

Base Pair	Combination	Base Pair	Combination
A·T	**Pu + Ac** Ac + AM Ac + At	T·A	**Gt' + Py** Me + Gt' At + Gt'
$^{me}A·T$	MA	T·^{me}A	MT
A·T or $^{me}A·T$	Pu + Ac' Pu + At Ac' + AM Ac' + At AM + At	T·A or T·^{me}A	Gt + Py At + Py Me + Gt At + Gt At + Me
G·C	GV	G·G	CU
G·^{me}C	Gt + AM Gt' + AM Gc + Am	^{me}C·G	Me + Ac' Me + Ac Me + Gc
G·C or G·^{me}C	**Pu + Gt** Pu + Gt' Pu + Gc Gt' + Gc Gt + Gc	C·G or ^{me}C·G	**Ac' + Py** Ac + Py Gc + Py Gc + Ac Gc + Ac'

Unambiguous recognition of base pairs requires at least two protein-base interactions, which would be pairings of the interactions listed in Table 2. If all physically possible combinations of such pairings are examined, they fall into three categories: Those that cannot be satisfied by any base pair, those that are still not unique and those that unambiguously specify a single base pair, which are shown here. The methylation states of adenine and cytosine are differentiated by some combinations, while others are insensitive to this modification. These are differentiated in the table. The combinations actually observed in *Eco*R I endonuclease are shown in bold face type. It should be noted that this table applies only to DNA.

A Mechanism for EcoR I* Activity

Any attempt to provide a recognition mechanism for EcoR I endonuclease must also account for the fact that the extraordinarily high cleavage specificity under physiological conditions can be relaxed by simple buffer conditions. In the altered conditions, EcoR I endonuclease recognizes many nucleotide sequences that differ from the canonical site, GAATTC, at one or more base pairs [29, 30]. The altered buffer conditions include; elevated pH 98 to 9.5), substituting Mn^{2+} for Mg^{2+}, low ionic strength, and the addition of organic compounds such as glycerol or etheylene glycol. The modified sequences, termed EcoR I* sites, are cleaved at variable rates which can be summarized by the simple hierarchial rules: G>>A>T>>C at the first position (that is, GAATTC is cleaved much faster than AAATTC, which is cleaved slightly faster than TAATTC, which in turn is cleaved much faster than CAATTC). Similarly the hierarchy at the second and third positions is A>>[G,C]>>T [31].

The 12 hydrogen bonds are consistent with these EcoR I* hierarchies if it is assumed that the protein adjusts its structure in order to maintain as many of the protein-base hydrogen bonds as possible. For example, if adenine were substituted for guanine at the first position, at least one hydrogen bond would be ruptured (the one in W2) as we have seen. Similarly, thymine could form, at most, one hydrogen bond with Arg^{200} because thymine has a methyl group in W1 (and an acceptor in W2). Thus, the EcoR I* sequences AAATTC and TAATTC could form, at most, 11 protein-base hydrogen bonds (one at the first position and two at each of the succeeding five canonical positions). Cytosine could not form any hydrogen bonds with Arg^{200} because it does not have any hydrogen bond acceptors in the major groove. Thus, the sequence CAATTC could form only ten sequence specific hydrogen bonds. If the bases in the first position of these hexanucleotides are ordered by the total number of protein-base hydrogen bonds, we obtain the sequence G, (A, T), C; the observed EcoR I* hierarchy.

The observed EcoR I* hierarchy is also obtained at the position of the second base; GGATTC has 11 possible protein-base hydrogen bonds with the loss of the hydrogen bond in W2 where a hydrogen bond donor on adenine is "converted" to a receptor on guanine. GCATTC also has 11 possible protein-base hydrogen bonds because of the loss of one in W1 where an acceptor N-7 on adenine is "replaced" with the hydrogen atom on the C-5 of cytosine. GTATTC has ten possible hydrogen bonds because no hydrogen bonds can be formed between the protein and the thymine at the second position. This count gives that sequence A, (G, C), T, the observed EcoR I* hierarchy. The identical result is obtained at the third position. The fourth, fifth, and sixth positions follow from the symmetry of the EcoR I site.

Thus all the EcoR I* hierarchies can be correlated with the maximum number of posible hydrogen bonds between the enzyme and the particular EcoR I* sequence. The idea that the number of protein-base hydrogen bonds determines the EcoR I* hierarchies was first suggested by Rosenberg and Greene [31], who correctly identified the major groove contacts subsequently observed in the electron density map. The hydrogen bonds counted in this article do not include numerous hydrogen bonds between the protein and the DNA backbone because they should not be affected by base substitutions.

Electrostatic Interactions and Recognition

The amino acid residues that hydrogen bond to the bases are electrically charged and they are arranged in space such that adjacent groups are oppositely charged. The four amino acid residues in the inner recognition module are located around the central twofold crystallographic axis (Fig. 6); the residues from the outer recognition modules, Arg^{200} and Arg^{203}, are above and below that axis. When the DNA phosphates are included in the charge distribution, there is alteration of charges over the entire complementary binding site forming a very stable array of electrostatic charges.

The negative charges associated with the carboxyl groups of Glu^{144} (from both subunits) are "keystones" of the electrostatic array. It is likely that a significant displacement of either or both of these side chains would lead to disruption of the entire recognition interface. The Glu^{144} interacts with the central adenine bases at the site where the EcoR I methylase modifies the DNA; that is, at the exocyclic N-6 amino group. Methylation of either or both of these sites would inevitably displace one or both Glu^{144} side chains. If the endonuclease were to bind to a methylated EcoR I site, then the direct hydrogen bonds from N-6 would be lost and the electrostatic character of the interface would be destabilized. Thus, the EcoR I endonuclease recognition interface seems highly poised to discriminate between the modified and unmodified hexanucleotides.

The spatial alternation of electrostatic charge suggests that the protein-DNA binding energy is a nonadditive function of the number of hydrogen bonds between the protein and the DNA bases; this means that each "correct" interaction should facilitate the formation of additional interactions of DNA and protein via the electrostatic forces. However, an "incorrect" structure due to the presence of a noncognate base in the enzyme's recognition site, would not facilitate and may even inhibit formation of additional protein-base interactions. These electrostatic interactions therefore constitute a form of cooperativity that would serve to sharpen the discrimination between the canonical hexanucleotide and all the incorrect sequences. We refer to this phenomenon as cooperative enhancement of specificity. Cooperative enhancement is also suggested by bonding data with oligonucleotide substrates which show that the bonding free energy is not a linear sum over the available hydrogen bonding sites [32]. Nonadditivity has also been observed in the interaction between the lac repressor and operator [33], which could represent a second example of cooperative enhancement.

Conformational Change and Specificity

From a mechanistic viewpoint, the difficult theoretical problem is not to "explain" the EcoR I* activity; rather it is to understand the physical basis of the highly precise canonical specificity that occurs at physiological conditions. The hierarchial spectrum of EcoR I* sites is just what we should expect from a simple energy analysis of a recognition mechanism based solely on hydrogen bonds. Loss of a single hydrogen bond would be expected to reduce the interaction energy by 1 to 4 kcal/mol. The energy would probably be reduced further by an additional term due to cooperative enhancement. The resulting reduction in association constant or catalytic rate constant would be about two to four orders of magnitude. In other words, a recognition mechanism based solely on binding and hydrogen bonds would predict an "error rate" that is comparable to the misreading associated with EcoR I* activity. However, under physiological conditions, there is no detectable activity at EcoR I* sites.

The amino acid side chains that interact with the specific bases do not participate directly in the cleavage reaction and vice versa because the recognition and cleavage sites are physically separate. It is not an accident of the crystalization procedure that the cleavage site is not assembled in the structure reported here. Rather, our current working hypothesis is that this structure represents a functional intermediate in the catalytic pathway. Specifically, we propose that the recognition and cleavage sites are formed in an obligate temporal order that includes an isomerization from an "inactive" form to an active form of the sequence-specific complex. The structure reported here is the specificially bound inactive conformer. Furthermore, we suggest that there is physical coupling between the recognition and cleavage sites. As a result, the enzyme retains the inactive conformation under physiological conditions until all the sequence specific DNA-protein interactions have formed. The transition is a form of allostery since the recognition and cleavage sites are spatially separate. We refer to the sequence dependent, allosteric isomerization from the inactive to the active form as "allosteric activation". Part of the free energy obtained from binding Mg^{2+} may be used to augment the sequence specificity of allosteric activation. Relatively subtle effects could dramatically alter the equilibrium between the inactive and active states. It is not unreasonable to argue that EcoR I* buffer conditions alter that equilibrium toward the active form even when one or two hydrogen bonds have not formed correctly.

Modrich and co-workers have independently arrived at the allosteric activation hypothesis in order to account for two observations they have recently made [34]. (i). Their kinetic data show that during the normal catalytic cycle, EcoR I endonuclease is bound to nonspecific DNA (which is not hydrolyzed). The data show that the total lifetime of all bound states is not the determinant of the cleavage rate; and they show that there are multiple bound states, some of which are inactive for cleavage. (ii) A mutation that replaces Glu^{111} with Gly retains full DNA binding specificity, but shows no cleavage activity under physiological conditions. Under EcoR I* conditions, the mutant enzyme cleaves DNA at EcoR I sites (at a rate much slower than that of wild type). Modrich and co-workers suggest that the mutation interferes with an isomerization between an inactive and active form. Glu^{111} is not near the DNA and cannot directly participate in the formation of either the recognitionor cleavage sites.

The very high sequence specificity seen in EcoR I endonuclease derives from a series of sequence specific steps including DNA binding and allosteric activation. Errors are corrected at each step via dissociation of noncognate DNA-protein complexes, resulting in a very low rate of cleavage at noncognate sites. This analysis suggests that an enzyme that covalently modifies DNA is intrinsically capable of achieving a much higher level of sequence discrimination than is a simple binding protein. Thus, sequence specific covalent modification of DNA may be important in higher organisms, which contain large quantities of DNA and which must precisely regulate crucial cellular events, such as those associated with development.

RECAPITULATION

The EcoR I endonuclease-DNA recognition complex consists of a distorted double helix and a protein dimer composed of identical subunits related by a twofold axis of rotational symmetry. The distortions of the DNA are induced by the binding of the protein. They are concentrated into separate features that are localized disruptions of the double helical symmetry. These disruptions appear to have structural consequences that propagate over long distances through the DNA via twisting and perhaps bending effects. They are therefore referred to as neokinks. The type I neokink spans the central twofold symmetry axis of the complex, and it introduces a net unwinding of 25° into the DNA. The unwinding increases the separation of the DNA backbones across the major groove thereby facilitating access by the protein to the base edges, which are at the floor of the groove. The type I neokink also realigns adjacent adenine residues within the central AATT tetranucleotide in order to create the detailed geometry necessary for amino acid side chains to bridge across these purines.

Each protein subunit is composed of a single principal domain with a central five-stranded wall of β sheet bracketed by α helices; that is, it is organized according to α/β architecture. Each domain also has an extension called an arm, which wraps around the DNA. The domain can be subdivided into topological motifs that have identifiable functional roles. The three-stranded parallel motif is associated with sequence recognition and the subunit interface. The three-stranded antiparallel motif is associated with phosphodiester bond cleavage. The two segments overlap to form the five-stranded β sheet.

The surface of the protein is involuted to form two symmetry related clefts which bind segments of the DNA backbone including the scissile bond. The cocrystals were grown in the absence of Mg^{2+}, in order to prevent DNA cleavage, but they can be activated for strand scission by diffusing Mg^{2+} into the crystals. The structure reported in this article appears to represent a specifically bound, inactive conformer that isomerizes to a specifically bound, active enzyme upon addition of Mg^{2+}. We suggest that the isomerization plays the important functional role of enhancing the specificity of EcoR I endonuclease by allosteric activation. The protein-base interactions at the sequence recognition site have a strong allosteric effect on the equilibrium between the inactive and active forms so that the active form is favored only when the cognate sequence is bound (under physiological conditions). The allosteric activation model accounts for the relaxation of specificity under EcoR I* conditions by invoking a solvent-mediated shift of the conformational equilibrium toward the active form even when EcoR I* sites are bound to the protein.

Sequence specificity is mediated by 12 hydrogen bonds between the protein and bases within the EcoR I hexanucleotide. These interactions depend on both the relative positioning as well as the identity of the bases and amino acid side chains at the DNA-protein interface. Unitary α helices position the key amino acid residues with respect to the DNA. These α helices are organized into modules with a spatial division of labor across the recognition site. The outer G·C base pairs are recognized by identical, symmetry-related outer modules. Each outer module consists of a single α helix. The inner tetranucleotide, AATT, is recognized by an inner module which consists of two symmetry related α helices, one from each subunit. Amino acid side chains from the modules establish the relative position of the α helices so as to form a four-helix bundle. Additional amino acid side chains position the bundle with respect to the DNA by interacting with the DNA backbone and by anchoring the recognition bondle within secondary structure of the complex.

Bidentate hydrogen bonds between Arg^{200} and guanine (Arg::G) determine the base specificity of the outer module. Substitution of any base other than guanine would lead to rupture of at least one of these hydrogen bonds. The inner module also utilizes bidentate hydrogen bonds, but in a bridging tetrad arrangement with Glu^{144} and Arg^{145} forming four hydrogen bonds to adjacent adenine residues (Glu-Arg::AA). Substitution of any other base for either adenine residue would also result in rupture of at lease one hydrogen bond. No hydrogen bonds are formed with the pyrimidine residues; however they are recognized by hydrogen bonds to the purines on the complementary strand. The 12 hydrogen bonds therefore occur only in the canonical EcoR I hexanucleotide. These interactions are also consistent with the spectrum of EcoR I* cleavage rages bacause the observed hierarchies of cleavage rates can be predicted simply by counting the maximal number of hydrogen bonds possible between the protein and relevant EcoR I* sites.

The recognition interactions are stabilized by interactions between amino acid side chains, including electrostatic interactions between oppositely charged pairs; Glu^{144}-Arg^{145} and Glu^{144}-Arg^{200}. These interactions suggest that the DNA-protein interaction energy is not a simple additive sum over the individual interactions; that is, the system utilizes cooperative enhancement to sharpen the discrimination between cognate and noncognate sites. Formation of some correct protein-base interactions facilitates formation of additional correct interactions whereas incorrect interactions with non-cognate bases have an inhibitory effect. Glu^{144} side chains from both subunits are centrally located in the electrostatic array. Methylation of either N-6 amino group by EcoR I methylase would rupture a hydrogen bond and displace one of these negative charges. The charge displacement should perturb the entire recognition interface, thereby sharpening the discrimination between the modified and unmodified EcoR I sites.

ACKNOWLEDGEMENTS

We thank Paul Modrich for sharing unpublished results, William Provost for his assistance with PS340 raster graphics and Oliver Bashor for technical assistance. This work was supported by NIH grant GM25671 (JMR). Additional support was derived from BRSG Grant RR07084, GM3306 (HWB) and GM25729 (PG).

REFERENCES

1. Greene, P.J., Gupta, M., Boyer, H.W., Brown, W.E. and Rosenberg, J.M., J. Biol. Chem., (1981) 256: 2143-2153.
2. Newman, A.K., Rubin, R.A., Kim, S.-H. and Modrich, P., J. Biol. Chem., (1981) 256: 2131-2139.
3. Modrich, P. and Zabel, D., J. Biol. Chem, (1976) 251: 5866-5874.
4. Jen-Jacobson, L., Kurpiewski, M., Lesser, D., Grable, J., Boyer, H.W., Rosenberg, J.M. and Greene, P.J., J. Biol. Chem., (1983) 258: 14638-14646.
5. Connolly, B.A., Eckstein, F. and Pingoud, A., J. Biol. Chem. (1984) 259: 10760-10763.
6. Modrich, P., Quart. Rev. Biophys., (1979) 12: 315-369.
7. Halvord, S.E. and Johnson, N.P., Biochem J., (1980) 191: 593-604.
8. Rosenberg, J.M., Boyer, H.W., and Greene, P.J., in Gene Amplification and Analysis Volume I: Restriction Endunucleases, J.G. Chirikjian, Ed. Elsevier/North-Holland, 1981, pp. 131-164.
9. Jack, W.E., Rubin, R.A., Newman, A. and Modrich, P., in Gene Amplification and Analysis Volume I: Restriction Endonucleases, J.G. Chirikjian, ed., Elsevier/North Holland, 1981, pp. 165-179.
10. Woodhead, J.L. and Malcolm, A.D.B., Nucl. Acids Res., (1980) 8: 389-402.
11. Modrich, P., CRC Crit. Rev. Biochem., (1982) 13: 287-323.
12. Terry, B.J., Jack, W.E., Rubin, R.A., and Modrich, P., J. Biol. Chem. (1983) 258: 9820-9825.
13. Ehbrecht, H.-J., Pingoud, A., Urbanke, C., Maass, G., and Gualerzi, C., J. Biol. Chem. (1985) 260: 6160-6166.
14. Terry, B.J., Jack, W.E. and Modrich, P., J. Biol. Chem. (1985) -- Submitted.
15. Grable, J. Frederick, C.A., Samudzi, C. Jen-Jacobson, L. Lesser, D., Greene, P.J., Boyer, H.W., Itakura, K. and Rosenberg, J.M., Journal of Biomolecular Structure and Dynamics (1984) 1: 1149-1160.
16. Frederick, C.A., Grable, J., Melia, M., Samudzi, C., Jen-Jacobson, L., Wang, B.-C., Greene, P.J., Boyer, H.W. and Rosenberg, J.M., Nature (1984) 309: 327-331.
17. McClarin, J.A., Frederick, C.A., Wang, B.-C., Greene, P., Boyer, H.W., Grable, J., and Rosenberg, J.M., Science (1986) 234: 1526-1541.
18. Kim, R., Modrich, P. and Kim, S.-H., Nucl. Acids Res. (1984) 12: 7285-7292.
19. Rosenberg, J.M., Seeman, N.C., Day, R.O., and Rich, A., Biochem. Biophys. Res. Commun., (1976) 69: 979-987.

20. Richardson, J.S., Adv. in Protein Chem.,(1981) 34: 167-339.
21. Rossmann, M.G., Liljas, A., Branden, C-I., and Banaszak, L.J., in The Enzymes, Boyer, P., ed, ?, Vol. 11, (1975) pp. 61-102.
22. Hol, W.G.S., Prog. Biophys. Molec. Biol., (1985) 45: 149-195.
23. Picone, J., Kim, Y., McClarin, J.A., Greene, P., and Rosenberg, J.M., in preparation.
24. Lu, A-L., Jack, W.E., and Modrich, P., J. Biol. Chem. (1981) 256: 13200-13206.
25. Lillehaug, J.R., Kleppe, R.K. and Kleppe, K. Biochemistry (1976) 15: 1858-1865.
26. Greene, P.J., Poonian, M.S., Nussbaum, A.L., Tobias, L., Garfin, D.E., Boyer, H.W. and Goodman, H.M., J. Mol. Biol., (1975) 99: 237-261.
27. Seeman, N.C., Rosenberg, J.M. and Rich, A., Proc. Natl. Acad. Sci. USA, (1976) 73: 804-808.
28. Rosenberg, J.M., unpublished.
29. Polisky, B., Greene, P., Garfin, D.E., McCarthy, B.J., Goodman, H.M. and Boyer, H.W., Proc. Natl. Acad. Sci. USA, (1975) 72: 3310-3314.
30. Woodhead, J.L., Bhave, N. and Malcolm, A.D.B., Eur. J. Biochem. (1981) 115: 293-296.
31. Rosenberg, J.M. and Greene, P.J. DNA (1982) 1: 117-124.
32. Jen-Jacobson, L., Lesser, D., and Kurpiewski, M., Cell (1986) 45: 619-629.
33. Mossing, M.C., and Record, M.T. Jr., J. Mol. Biol. (1985) 186: 295-305.
34. Modrich, P., personal communication.
35. Dickerson, R.E., and Drew, H.R., J. Mol. BIol. (1981) 149: 761-786.
36. Dickerson, R.E., J. Mol. Biol. (1983) 166: 419-441.
37. Dickerson, R.E., and Drew, H.R., Proc. Natl. Acad. Sci. USA (1981) 78: 7318-7322.

5

The Enzymes of the *Bam*H I Restriction-Modification System

Glenn Nardone and Jack G. Chirikjian

Georgetown University
Department of Biochemistry
Lombardi Cancer Research Center
3800 Reservoir Road
Washington, DC 20007

I. INTRODUCTION

The remarkable sequence specificity of Type II restriction endonucleases and their cognate methyl transferases has recently placed them under the scrutiny of nucleic acid enzymology. From a biochemical perspective, their comparatively uncomplicated reaction requirements, intermediate size and concise, well-defined recognition sites has made them attractive models for the investigation of specific protein-nucleic acid interactions. The solution of the same DNA sequence recognition problem by the enzymes of a restriction-modification system has lead to interesting questions concerning structure-function relationships between these genetically distinct proteins. Further study of restriction-modification enzymes should promote a better understanding of the patterns and principles of protein-nucleic acid interactions.

The type II restriction-modification enzymes of *Bacillus amyloliquefaciens* H recognizes the duplex, symmetrical sequence 5'-GGATCC-3' [1,2]. In the presence of Mg^{+2} the endonuclease catalyzes double stranded cleavage between the guanines, generating 5'-phosphoryl and 3'-hydroxyl staggered termini. The methylase catalyzes methyl group transfer from S-adenosyl-L-methionine to the C^5 position of the internal cytosines [3]. Methylation prevents cleavage by the endonuclease and is the presumed host controlled mechanism for the protection of endogenous DNA. The specificity of the methyl acceptor must be as stringent as the position of strand scission since methylation of the external cytosines or the 6-amino groups of the adenines does not prevent cleavage [4,5]. We have been involved with the purification and characterization of these enzymes. Emphasis has been placed on their catalytic properties and mechanisms of sequence discrimination.

II. PURIFICATION AND STRUCTURAL CHARACTERISTICS

    A. Purification of *Bam*H I endonuclease

*Bam*H I endonuclease has been purified to apparent homogeneity using the procedure summarized in Table I [6]. All procedures were carried out at 4°C. In a typical preparation frozen *Bacillus amyloliquefaciens* cells (RUB 500) were thawed in 20 mM phosphate (K^+), pH 7.2, 10 mM 2-mercaptoethanol and 1 mM Na_2EDTA (buffer A). The cells were homogenized in a Waring blender, filtered through cheesecloth and lysed by three passages through a Gaulin homogenizer at 9000 psi. Cell debris was removed by centrifugation at 8000 x g for 30 minutes. In small scale preparations (500 grams of cells) subcellular material and particulates were removed by centrifugation at 100,000 x g for 30 minutes. Preparations involving 3 to 6 kilograms of cells have been cleared of subcellular material by filtration under nitrogen through 142 mm Millipore filter units employing various grades of prefilters and 3 micron membranes.

TABLE I. Purification of Bam HI Endonuclease from 0.5 Kg. of B. amyloliquefaciens

Fractionation Step	Total Protein mg.	Total Units (x 10^{-4})	Specific Activity[a] units/mg	Yield %	Purity -fold
Crude Extract	18,800	11.2	6	100	1
Phosphocellulose	320	6.9	215	62	36
Hydroxyapatite	40	6.4	1,600	57	267
Bio-Rex 70	5	2.4	4,800	21	800
Aminopentyl-Sepharose	2	2.1	10,500	19	1750

[a]Units were determined with form I pMB9 DNA (one Bam HI site). One unit of endonuclease will cleave 1 pmol of phophodiester bonds per min. at 37°C in 25 mM Tris-HCl, pH 8.5, 2 mM 2-mercaptoethanol, 10 mM $MgCl_2$.

The crude extract was applied to a phophocellulose column equilibrated in buffer A. The column was washed until the effluent was less than 0.05 at 280 nm. The column was then developed with a linear gradient of NaCl from 0 to 0.8 M. Peak activity eluted between 0.1 and 0.35 M NaCl. Enzyme fractions were applied to hydroxyapatite column equilibrated in buffer A. After washing the column was developed with a 0.02 to 0.3 M gradient of phosphate (K^+), pH 7.2. Endonuclease activity eluted between 0.07 and 0.11 M phosphate. Enzyme fractions were pooled and dialyzed against 20 mM Tris-HCl[a], pH 7.5, 1 mM Na_2EDTA and 2 mM 2-mercaptoethanol (buffer B). The enzyme was then applied to a BioRex 70 column equilibrated in buffer B. After the removal of an inert protein peak during washing the enzyme was eluted with a 0 to 0.5 M NaCl gradient. Peak endonuclease activity occurred between 0.2 and 0.4 M NaCl. Following dialysis against buffer B the enzyme was applied to an aminopentyl sepharose column. The column was developed with a 0 to 0.6 M NaCl gradient. Peak activity was found between 0.2 and 0.3 M NaCl. Aliquots of column fractions were submitted to SDS-polyacrylamide electrophoresis and those fractions containing one band were pooled and desalted

[a]Abbreviations used in the text are Tris, tris-(hydroxymethylamino)methane; AdoHcy, S-adenosylhomocysteine; AdoMet, S-Adenosyl-L-methionine; BSA, bovine serum albumin; Na_2EDTA, disodium ethylenediamine tetraacetic acid; form I, supercoiled DNA; form II, open circular DNA; form III, full length linear DNA; RF, replicative form; SDS, sodium dodecyl sulfate; SV40, Simian virus 40 DNA; Tween 20, polyoxyethylene sorbitan monolaurate.

by dialysis or ultrafiltration with buffer B. Enzyme prepared by this method was free of non-specific exo and endonucleases. The protein was homogenous as judged by SDS-polyacrylamide gels stained with silver nitrate or Coomassie brilliant blue and analytical ultracentrifugation using ultraviolet optics.

B. Purification of BamH I Methylase

BamH I methylase has been purified to apparent homogeneity using the recently developed procedure outlined in Table II [7]. Methods of cell lysis and phophocellulose chromatography were the same as those described for the endonuclease. Both enzymes were separated by chromatography on phosphocellulose. Peak methylase activity eluted between 0.4 and 0.55 M NaCl. Enzyme fractions were pooled, made 0.85 M in ammonium sulfate (the pH was held constant by the addition of NaOH), filtered through 3 micron membranes, and applied to a phenyl sepharose column equilibrated in 20 mM phosphate (K^+), pH 7.2, 10 mM 2-mercaptoethanol, 1 mM Na_2EDTA and 0.85 M ammonium sulfate. The column was washed with buffer containing 0.65 M ammonium sulfate and was developed with a decreasing of 0.65 to 0 M ammonium sulfate. Methylase activity eluted between 0.4 and 0.2 M ammonium sulfate. The enzyme fractions were pooled and applied to a hydroxyapatite column equilibrated in buffer A. The column was developed with a linear gradient of 0.02 to 0.6 M phosphate (K^+), pH 7.2. Methylase eluted between 0.28 and 0.35 M phosphate. Enzyme fractions containing a single band as judged by SDS-polyacrylamide gels stained with Coomassie brilliant blue or silver nitrate were pooled and concentrated to 0.2 mg/ml by vacuum dialysis against buffer A. In some preparations a small amount of contaminating protein copurified. This was removed by chromatography on Cibracon blue agarose developed with a 0.2 to 1.5 M NaCl gradient. Methylase activity eluted between 0.7 to 1 M NaCl. Enzyme activity was stable for 1 to 3 months at $4^{\circ}C$. Storage life at these temperatures was increased in the presence of 0.15 M ammonium sulfate. Dialysis against 20 mM Tris-HCl, pH 7.5, 10 mM 2-mercaptoethanol, 1 mM Na_2EDTA, 50% glycerol and storage at $-20^{\circ}C$ preserved enzyme activity for 12 months. The methylase preparation was free of contaminating BamH I endonuclease and non-specific nucleases. Incubation of excess enzyme with Col EI or ϕX174 RF DNAs (both do not contain a BamH I site) did not result in significant amounts of methyl group incorporation. Functional purity was also assessed with pUC9 DNA (contains one BamH I site) previously cleaved with BamH I endonuclease (Figure 1). No methyl group incorporation was observed but near stoichiometric incorporation occurred with form I pUC9 DNA.

TABLE II. Purification of Bam HI Methylase from 4 Kg. of
B. amyloliquefaciens

Fractionation Step	Total Protein mg.	Total Units (x 10^{-4})	Specific Activity[a] units/mg	Yield %	Purity -fold
Crude Lysate[b]	110,000	-	-	-	-
Phosphocellulose	1,500	750	500	100	-
Phenyl-Sepharose	18	450	25,000	60	50
Hydroxyapatite	1.6	305	197,000	45	394

[a]Units were determined with lambda DNA (five Bam HI sites). One unit of methylase will protect 1 ug of lambda DNA from cleavage by excess Bam HI endonuclease in 1 hour at 37°C in 50 ul reactions containing 25 mM phosphate (K^+), pH 7.2, 5 mM Na_2EDTA, 5 mM 2-mercaptoethanol, 100 mM KCl, 200 ug/ml BSA, 10 uM AdoMet.

[b]Units could not be determined because of dilute methylase concentrations and inhibition from contaminants.

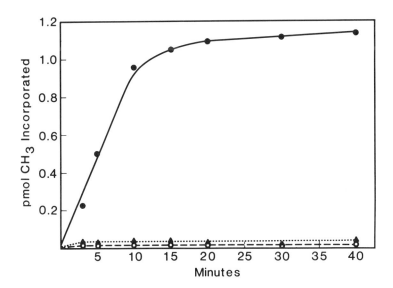

Figure 1: Functional Purity of Bam HI Methylase

Bam HI methylase was incubated with form I pUC9 DNA (●), pUC9 DNA previously cleaved with Bam HI endonuclease (▲) or form I Col EI DNA (O). pUC9 DNA contains one Bam HI site. The concentration of DNA and ^{3}H-AdoMet were 12.2 nM and 10 uM respectively.

The methyl acceptor for purified BamH I methylase was verified by using Sau 3A endonuclease. The endonuclease recognizes 5'-GATC-3'[8,9] which is a subset of the BamH I sequence. Sau 3A endonuclease is unaffected by adenine methylation but cannot cleave when cytosine is methylated [10,11]. A 376 base pair fragment was prepared from pBR322 DNA. This fragment contained one BamH I site and three Sau 3A sites. Sau 3A cleavage generated 159, 99, 91 and 27 base pair fragments. BamH I methylation of the 376 base pair substrate and subsequent cleavage with Sau 3A resulted in 159, 118 and 99 base pair fragments (Figure 2). This is the expected cleavage pattern if the BamH I site is methylated at the internal cytosines.

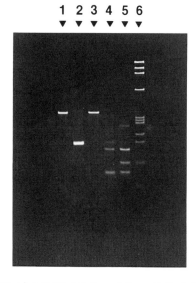

Figure 2: <u>Specificity of Bam HI Catalyzed Base Methylation as Judged by Resistance to Sau 3a Endonuclease Cleavage.</u>

The substrate DNA was a 376 base pair fragment of pBR322 DNA containing one Bam HI site and three Sau 3A sites. Reaction products were fractionated by electrophoresis in a 8% polyacrylamide gel.

Lane 1   Intact fragment
Lane 2   Bam HI cleaved fragment. The 190 and 186 base pair products migrate as a doublet.
Lane 3   Bam HI methylated fragment exposed to an excess of Bam HI endonuclease for 2 h.
Lane 4   Sau 3a cleaved fragment. The 99 and 91 base pair products migrate as a doublet. The 27 base pair fragment is not visible in this photograph.
Lane 5   Bam HI methylated fragment exposed to excess Sau 3a endonuclease for 2 h.
Lane 6   Hae III digest of phiXRF174 DNA. Fragment sizes, in base pairs, are 1353, 1078, 872, 603, 310, 281, 271, 234, 194, 118, 72.

C. Structural Characteristics

Gel filtration of BamH I endonuclease at ionic strengths up to 0.3 and protein concentrations of 0.1 mg/ml indicated a native molecular weight of 90,000. In the presence of 0.5 M NaCl the molecular weight was estimated to be 46,000 by gel filtration and sucrose density ultracentrifugation. SDS-polyacrylamide electrophoresis reveals a single band with a molecular weight of 22,000. These results suggest that the endonuclease consists of identical polypeptide chains which aggregate to dimers or tetramers depending on the ionic strength. Similar characteristics have been observed with EcoR I endonuclease [12]. However, the active form of EcoR I endonuclease appears to be the dimer as judged by gel filtration of enzyme-substrate complexes [13]. The ultraviolet spectrum of BamH I endonuclease is that of a typical protein. Isoelectric focusing indicated a pI of 6.7.

The N-terminal 40 amino acid sequence of BamH I endonuclease has been determined (Figure 3). Secondary structure predictions using the method of Chou and Fasman [14] were made for the N-terminal portions of BamH I and EcoR I endonucleases (Figure 4A). Some striking similarities were found. Beta turns were found to occur at positions 16-19 in both enzymes. Similarly, the analysis predicted runs of alpha-helix flanking both sides of the turn. Secondary structure predictions by the semithermodynamic approach of Rose [15] was also used. The relative hydrophobicity was plotted as a function of the residue number (Figure 4B). Minima in hydophobicity have been shown to correlate 95% with turns in secondary structure. Maxima are thought to be positions where the polypeptide backbone enters deeply into the internal portions of the protein. Although the hydrophobicity pattern for the N-terminal fragments of BamH I and EcoR I endonucleases are different, the minima, with one exception, occur within two amino acids of each other.

```
                                           10
        Met Glu Val Glu Lys Glu Phe Ile Thr Asp
                                           20
        Glu Ala Lys Glu Leu Leu Ser Lys Asp Lys
                                           30
        Leu Ile Gln Gln Ala Tyr Asn Glu  ?  Lys
                                           40
        Thr Ser Ile Ser Ser Pro Ile Trp Pro Ala
```

Figure 3: <u>The N-terminal 40 Amino Acid Sequence of *Bam* HI Endonuclease</u>

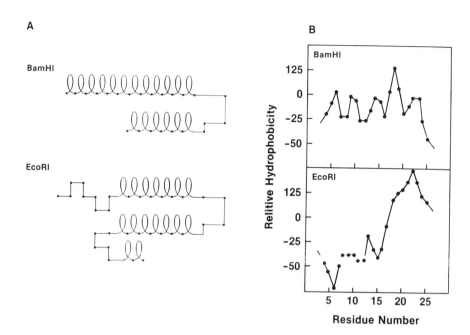

Figure 4: Secondary Structure Predictions for the N-terminal Region of *Bam* HI and *Eco* RI Endonuclease

    A. Secondary structure of *Bam* HI (upper) and *Eco* RI (lower) endonuclease as predicted by the method of Chou and Fasman.

    B. Secondary strucutre of *Bam* HI (upper) and *Eco* RI (lower) endonuclease as predicted by the semi-thermodynamic method of Rose.

SDS-polyacrylamide electrophoresis of *Bam*H I methylase revealed a single band with a molecular weight of 56,000. No change in migration was observed when 2-mercaptoethanol was eliminated from the denaturation steps prior to electrophoresis. Gel filtration and sucrose density ultracentrifugation indicated a native molecular weight of approximately 64,000 and a Stokes radius of 34 angstroms. A S_{w20} of 4.4 was calculated. Peak activity fractions from both experiments were subjected to SDS-polyacrylamide gel electrophoresis which again resulted in a single band having a molecular weight of 56,000 as visualized by silver nitrate or Coomassie stains. No change in the native molecular weight was observed in the presence of 0.02 - 1 M KCl, 1-35 uM AdoMet or at initial protein concentrations in the range of 0.01 to 0.5 mg/ml. These data indicate that the enzyme exists as a monomer.

III. CATALYTIC PROPERTIES

A. BamH I Endonuclease

BamH I endonuclease exhibits activity over a broad pH range, with an optimum at 8.5 in Tris buffers. The optimum Mg^{+2} concentration is 10 mM. The addition of Tween 20 or BSA to the reaction solution greatly stabilized enzyme activity at 37°C. The enzyme was most active at 37-40°C and inactivation occurred at 55°C in the absence of substrate or NaCl. Enzyme activity was stabilized to approximately 65°C in the presence of 100 mM NaCl. However, the enzyme was optimally active at lower salt (0-50 mM) and is severely inhibited at 250 mM NaCl. DNA stabilized the enzyme at 37°C and at dilute protein concentrations.

Initital velocity studies performed with form I SV40 or pBR322 DNAs (both contain one BamH I site) revealed hyperbolic kinetics with Kms in the vicinity of 3.6 nM. The estimated turnover number was 1.5 phophodiester bonds cleaved min^{-1}. The 5'phosphoryl deoxydinucleotide subsets of the BamH I sequence (GG, GA, AT, TC, CC) have been shown to be specific inhibitors of the endonuclease [16,17,18]. Reciprocal plots of representative reactions containing GG and CC indicated competitive inhibition patterns (Figure 5). Dinucleotides unrelated to the BamH I sequence such as AA, TT and GC did not inhibit the endonuclease. The K_i values of the dinucleotides varied over a wide range (Table III) suggesting differences in the interactions of the endonuclease at various points within the recognition sequence. This idea is supported by alkylation interference studies conducted with specific complexes between EcoR I endonuclease and DNA which demonstrated preferences in potential recognition contacts with the purines and phosphates of the sequence [19]. The dinucleotides CC and GG had the most potent K_{is}, suggesting that the ends of the recognition site are most important to the formation of a specific complex.

The inhibitory effects of the dinucleotides were found to be synergistic in mixing experiments (Table IV). This may be due to conformational changes in the enzyme caused by the binding of an inhibitor which affects the binding of another. Nitrocellulose filter binding experiments measuring the displacement of stable endonuclease-SV40 DNA complexes (formed in the absence of Mg^{+2}) by the dinucleotides were used to construct Hill plots [20]. Hill coefficients were >1 suggesting multiple interacting sites for the inhibitors. The existence of at least two binding sites in the active site for each inhibitor seems reasonable considering the symmetrical, double stranded BamH I site.

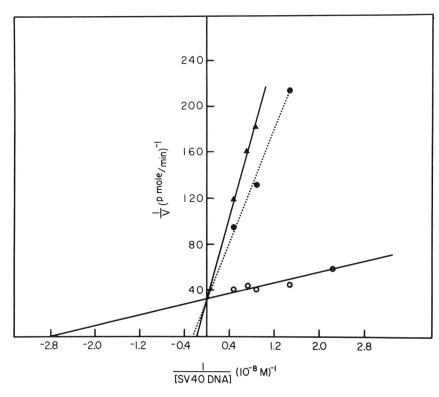

Figure 5: <u>Reciprocal Plots of the Initial Velocity of Bam HI Endonuclease in the presence of pdGpdG and pdCpdC</u>

Bam HI endonuclease (0.8 nM) was incubated with various concentrations of form I ^{3}H-SV40 DNA in the absence of dinucleotides (O), 45 uM pdGpdG (●) or 20 uM pdCpdC (▲) for 5 min. at 37°C. Form II DNA (specifically nicked at the Bam HI site) and form III DNA were equated to 1 and 2 phosphodiester bonds cleaved, respectively. The initial amount of randomly nicked form II DNA was approximately 4%.

TABLE III. Inhibition Constants of Bam HI Deoxydinucleotide Subsets with Bam HI Endonuclease

Dinucleotide	$K_I{}^a$
pdCpdC	1.2
pdGpdG	4.2
pdTpdC	58
pdApdT	117.6

$^a K_I$ values were determined by extrapolation on replots of the apparent K_M values <u>versus</u> the dinucleotide concentration.

TABLE IV. Synergistic Inhibitory Effects of the Deoxynucleotide Inhibitors on *Bam* HI Endonuclease Activity

Dinucleotide	uM	Remaining Activity %
GG	20	85
GA	20	100
TC	20	91.7
AT	20	96.4
CC	3	93.5
GG + GA	20, 20	75.5
GG + AT	20, 20	69.1
GG + CC	20, 3	42.8
GG + TC	20, 20	66.6
CC + GA	3, 20	81.9
CC + AT	3, 20	75.3
CC + TC	3, 20	61.9
GA + AT	20, 20	90.0
GA + TC	20, 20	70.9
AT + TC	20, 20	73.1
Control[a]	--	100

[a] 100% = 0.011 pmol phosphodiester bonds cleaved per min.

The kinetic mechanism of *Bam*H I endonuclease with form I SV40 DNA was examined in greater detail with time course experiments. Time course data was analyzed by the computer program of Berman [21]. This program employs a combination of numerical integration of differential equations in terms of rate constants which have been determined by coordinated non-linear least squares analysis of the data. Evaluation of the time course data was done in terms of the proposed kinetic model for *Eco*R I endonuclease [22].

$$E+S \underset{k_{-1}}{\overset{k_1}{\longleftrightarrow}} ES \overset{k_2}{\to} EN \overset{k_3}{\to} EL \overset{k_4}{\to} E + L$$

$$k_5 \updownarrow k_{-5}$$
$$E + N$$

$$K_N = k_5/k_{-5}$$

E = enzyme
S = form I DNA
N = form II DNA
L = form III DNA

The following assumptions were made to simplify the analysis: k_1 and k_{-1} were large with respect to k_2, i.e. the binding of superhelical DNA to the enzyme was in rapid equilibrium with respect to the first cleavage step; k_4 was not specified. The results of this analysis are shown in Figure 6. The values of k_2, k_3 and K_N were 0.192 min^{-1} ± 0.0019, 1.190 min^{-1} ± 0.042 and 1.07 ± 0.067 respectively. This approach promises to be powerful since a kinetic model can be quantitatively analyzed in terms of simultaneously determined, self-consistent rate constants. According to the analysis the first strand cleavage was substantially slower than the second. The opposite was found in the *Eco*R I endonuclease reaction with SV40 DNA [22].

Flanking sequences might influence the rates of individual cleavage events if one strand is preferentially nicked. Studies with synthetic substrates have shown that Hpa II endonuclease favored cleavage of the strand that was rich in pyrimadine flanking sequences [23]. Rate constants for strand cleavage events might also be affected by superhelicity or the 5'-phosphate and 3'-hydroxyl groups generated in the form II intermediate.

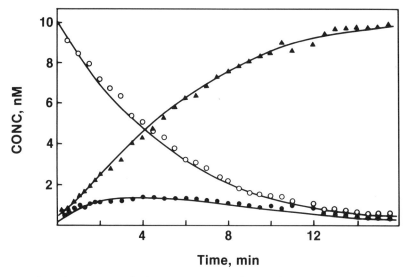

Figure 6. <u>Time Course of the Bam HI Endonuclease Cleavage of form I SV40 DNA</u>.

The cleavage of form I ^{3}H-SV40 DNA was followed in reactions containing 10 nM DNA and 2 nM endonuclease. Reaction cocktail containing enzyme, DNA and buffer was prewarmed at 37°C. for 5 min. The reaction was started by the addition of prewarmed MgCl$_2$ to a final concentration of 10 mM. Data analysis was conducted with computer fitting to the kinetic model described in the text. The lines represent the integrated solutions to the rate equations and the real data points are represented by form I DNA (O), form II DNA (●) and form III DNA (▲).

B. BamH I Methylase

BamH I methylase exhibited optimal activity in the presence of 100 mM NaCl or KCl. Salt concentrations in excess of 250 mM were severely inhibitory. The enzyme is inhibited at Mg^{+2} concentrations above 5 mM with complete inhibition occurring at 20 mM. Methylase activity was stabilized in the presence of thiols, BSA and Tween-20. AdoMet, AdoHcy or DNA increased the stability of dilute methylase at assay temperatures of 37-40°C. The pH profile of methylase activity (performed at a constant ionic strength of 0.1) was complicated, showing a maximum at pH 7.2 with a

secondary peak at pH 7.8. Multiple pH maxima have also been observed with
BsuR I methylase [24].

The reaction characteristics of the methylase and endonuclease were
examined with double stranded M13mp8 DNA containing a single hemimethylated
BamH I site. This substrate was prepared by methylating M13mp8 RF DNA with
an excess of BamH I methylase. The DNA was then linearized with Bal I
endonuclease and hybridized with a 10-fold excess of single stranded M13mp8
DNA (+ strand). Single stranded DNA was purified from form II hybrid DNA by
chromatography on hydroxyapatite. BamH I endonuclease did not catalyze
double stranded cleavage of hemimethylated DNA (Figure 7). In contrast,
methylation rates with hemimethylated substrate were approximately 3 to 4
fold greater than with unmethylated controls (Figure 8). These results may
have physiological significance. Semiconservative replication of DNA would
be expected to generate transiently hemimethylated recognition sites. These
sites are not substrates for resident restriction enzymes and therefore
obviates the need for elaborate regulation mechanisms to prevent cleavage.
Faster methylase kinetics at hemimethylated sites would favor the completion
of methylation of endogenous DNA before the next round of replication. Both
the endonuclease and methylase were inactive with a single stranded recognition site as determined with single stranded M13mp8 DNA.

The mechanism of BamH I methylation was evaluated with form I SV40 DNA
and catalytic amounts of enzyme. Plots of methyl group incorporation versus
percent BamH I cleavage generated straight lines with slopes of 50% (Figure
9). This data suggests that the enzyme transfers two methyl groups per DNA
binding event as opposed to successive transfers that are separated by
enzyme dissociation and reassociation steps with the DNA. However, this
data does not completely eliminate the latter mechanism. If the transfer of
the second methyl group is much more favorable than the first then a
straight line with a slope approaching 50% could result. In such a case the
line would represent a biased average of two separate methyl transfers. The
kinetic inequivalence of the methyl group transfers is supported by the data
in Figure 8. This does not necessarily mean that two transfers per binding
event cannot occur with unmethylated DNA because the enzyme may undergo a
mechanistic change that depends on the methyl group status of the initial
substrate. Such mechanistic cues apparently occur with the type I restriction enzyme Eco K. This enzyme switches from a cleavage to a highly
efficient methylation mode when its recognition site is hemimethylated [25].

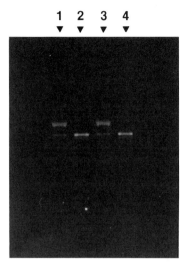

Figure 7: <u>Activity of *Bam* HI Endonuclease with Hemimethylated DNA</u>.

Excess *Bam* HI endonuclease was incubated with form II M13mp8 DNA hybrids for 1 h. Cleavage products and controls were fractionated on 0.8% agarose gels.

Lane 1	Unmethylated DNA control
Lane 2	Unmethylated DNA exposed to *Bam* HI endonuclease.
Lane 3	*Bam* HI hemimethylated DNA exposed to *Bam* HI endonuclease.
Lane 4	*Bam* HI hemimethylated DNA exposed to *Sal* I endonuclease. The *Sal* I recognition site is adjacent to the *Bam* HI site.

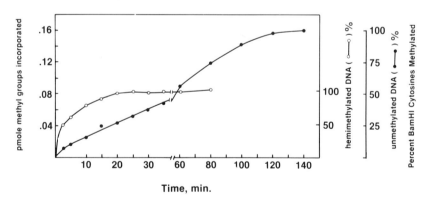

Figure 8: <u>Rates of Methylation with Unmethylated and *Bam* HI Hemimethylated DNA.</u>

Bam HI methylase was incubated in reactions containing 6 nM form II M13mp8 DNA (O) or *Bam* HI hemimethylated M13mp8 DNA (O) and 5 uM ^{3}H-AdoMet. M13mp8 DNA that was initially unmethylated had a final ^{3}H-methyl content of 2.03 moles per mole of DNA. M13mp8 DNA that was originally hemimethylated had a final ^{3}H-methyl content of 1.08 moles per mole of DNA. Both these values were normalized to 100% for this analysis.

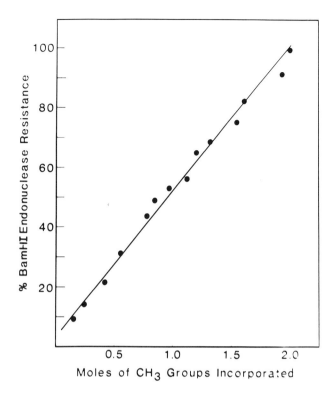

Figure 9: <u>Resistance of SV40 DNA to *Bam* HI Endonuclease as a Function of Methyl Group Content.</u>

Form I SV40 DNA was methylated in reactions containing 15 nM DNA, 5 uM ^{3}H-AdoMet and 0.8 nM *Bam* HI methylase over a period of 2 h. Aliquots were removed at various time intervals and made 25 mM in $MgCl_2$ to quench the reaction. The methyl group content of the DNA was determined and excess *Bam* HI endonuclease was added to the remainder of the sample. Cleavage products were fractionated by agarose gel electrophoresis and the percentage of nuclease resistant DNA was determined densitometrically. Completely protected DNA was found to have 2.1 moles of methyl groups per mole of DNA. This value was normalized to 2 for this analysis.

Initial velocity studies with *Bam*H I methylase revealed hyperbolic kinetics with form I SV40 and pBR322 DNAs. The apparent Km for the DNAs was approximately 4 nM in the presence of 35 uM AdoMet. The turnover number was estimated to be 1.3 methyl groups min^{-1}. The initial velocity of methylation was first order with respect to enzyme concentration in the range of 0.03 to 1.2 nM at saturating DNA concentrations of 30 nM. This result, in conjunction with gel filtration and centrifugation experiments performed at much higher protein concentrations, suggest that the methylase acts as a monomer. However, the aggregation state of the enzyme in the presence of

DNA is not yet known. Initial velocity kinetics with varying AdoMet at saturating and subsaturating DNA revealed a complex concentration dependence under a wide variety of conditions. The causes of this behavior with AdoMet are not yet clear. The kinetics become zero order at AdoMet concentrations around 26 uM. The possibility for multiple interactive sites of AdoMet has been considered although much more needs to be investigated in the areas of substrate binding and enzyme-substrate complexes that result in kinetically unproductive intermediates.

Kinetic studies performed with varying DNA and fixed-varying AdoMet generated a family of parallel lines on reciprocal plots. Since the methylase kinetics with AdoMet are apparently non-Michaelean more rigorous initial velocity studies could not be employed to ascertain if the enzyme operates by a ping-pong mechanism. In the absence of DNA, incubation of BamH I methylase with ^{14}C-AdoMet that was labelled in the alpha carboxylic group and unlabelled AdoHcy did not result in methyl group exchange to generate ^{14}C-AdoHcy. This ruled out the possibility of a kinetically significant methylated enzyme intermediate. The enzyme may form a covalent adduct at the C^6 position of the cytosine acceptors to produce an attacking C^5 carbanion species which becomes methylated. A mechanism such as this has been postulated for a t-RNA uracil methylase [26]. There are caveats to consider when interpreting kinetic plots having parallel lines. Apparent parallel lines may in fact be converging. Convergence may not be obvious because of errors in the velocity measurements or the size of the constant terms in the denominator of the rate equation [27]. If the dissociation constant of a substrate is much smaller than its Km parallel lines may be observed for an ordered mechanism in rapid equilibrium [28].

C. Kinetic Properties of BamH I Endonuclease and Methylase with Linear DNA Substrates

Non-specific flanking DNA is normally in great molar excess over specific recognition sequences. Non-specific sequences would be expected to increase the sampling time preceding the formation of a sequence specific complex between a protein and DNA. In addition, inhibitory competitive binding at non-specific sites (presumed to be primarily electrostatic in origin) can act to further reduce the rate of specific complex formation. However, non-specific flanking DNA appears to enhance the kinetic ability of lac repressor and EcoR I endonuclease to form sequence specific complexes beyond the predicted diffusion controlled limits [29,30]. These long range effects have been theoretically explained in terms of thermally driven facilitated diffusion mechanisms. These intramolecular transfer processes involve the sliding or hopping of a protein along the DNA contour length, dissociation-association events within a nucleotide domain and intersegment exchange of protein between areas of the DNA polymer brought into close proximity by conformational fluxes [31]. Such facilitated transfer mechanisms are expected to increase specific binding rate constants since they are inherently faster than random sampling by three dimensional diffusion.

The kinetics of BamH I endonulcease and methylase were examined with full length pBR322 DNA that had been linearized with different restriction enzymes. The endonuclease and methylase exhibited faster reaction rates with substrates having the recognition site in a more central position (Figure 10) [32]. Experiments with a variety of linearized pBR322 DNA substrates showed that these effects were correlated with the distance and not the direction (5' or 3') of the nearest DNA terminus to the BamH I site. Increasing NaCl concentrations decreased the differences between cleavage rates (Figures 10 B and C) while the differences in methylation rates remained relatively unaffected (Figures 10 E and F). The methylase did show its characteristic increase in overall activity in the presence of 100 mM NaCl (Figure 13 E). Initial velocity studies performed with both enzymes in the absence of NaCl demonstrated a 3 fold increase in the Km for EcoR I linearized pBR322 DNA over Nde I linearized pBR322 DNA (BamH I cleavage site is 375 and 1921 base pairs, respectively, from the nearest end). The methylase kinetics with AdoMet did not change when either DNA was used as the non-varying substrate.

The reaction rate differences described for the endonuclease and methylase could have originated from one or more types of facilitated diffusion. If these enzymes can locate their recognition site by a sliding or hopping mechanism then an enhancement of these processes would be expected to occur as the recognition site is located in a more central position. This would be a manifestation of the proportionately larger target area for non-specific binding around the BamH I site. Implicit in this assumption is that the average scanning length is less than 3987 and greater than 375 base pairs (the longest and shortest distances between a DNA terminus and the BamH I cleavage site on EcoR I linearized pBR322 DNA). A sliding mechanism has been implicated in the EcoR I endonuclease reaction as judged by increases in the specific association and dissociation rate constants with increasing DNA chain length up to approximately 4000 base pairs [30]. The average scanning length was determined to be 1300 base pairs and the reaction kinetics were faster with longer substrates. More recent work demonstrated that EcoR I endonuclease has a kinetic preference for a centrally positioned site and that this phenomenon can be explained by facilitated transfer [33].

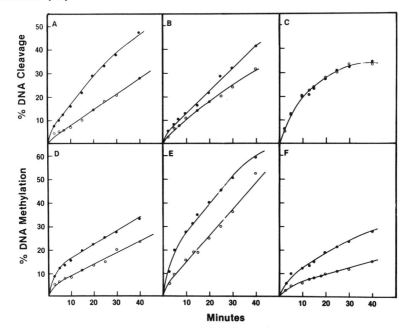

Figure 10: <u>Differences in Reaction Rates of <i>Bam</i> HI Endonuclease and Methylase as Related to the Relative Location of the <i>Bam</i> HI Site.</u>
The time course of cleavage and methylation was followed in reactions containing pBR322 DNA that had been linearized with <i>Eco</i> RI or <i>Nde</i> I endonuclease, placing the <i>Bam</i> HI cleavage site 375 (O) or 1921 (●) bases from the nearest end, respectively. The DNA concentration was 12 nM and the concentrations of endonuclease and methylase were 0.4 and 0.25 nM respectively. The AdoMet concentration was 35 uM. The rate of endonuclease cleavage under standard reaction conditions (A), in the presence of 100 mM NaCl (B), in the presence of 160 mM NaCl (C). The rate of methylation under standard reaction conditions (D), in the presence of 100 mM NaCl (E), in the presence of 200 mM NaCl (F).

In order to ascertain if facilitated diffusion is kinetically evident for the <i>Bam</i>H I enzymes fragments of pR322 DNA were prepared with the <i>Bam</i>H I site in a central location (Table V). Endonuclease cleavage rates increased 9 fold with increasing substrate length from 74 to 4362 base pairs (Figure 11). Differences between cleavage rates greatly reduced in the presence of 160 mM NaCl (Figure 11). In the absence of NaCl the kinetic preference for longer substrates was observed over a DNA concentration range of 0.1 to 12 nM. With the assumption that association was rate limiting the increases in cleavage rates could have been due to the larger target area for non-specific binding and, therefore, an increase in the frequency of facilitated transfer. The non-linear profile of cleavage rate increases indicates that the range of facilitated transfer is limited. The upper limit of the enzyme's scanning length (assuming a sliding or hopping mechanism) was calculated to be in the vicinity of 1300 base pairs from this data. The elimination of both long chain preference and the differences in cleavage rates between the full length substrates by high NaCl concentrations is consistent with a decrease in the electrostatic component of non-specific binding.

TABLE V. Fragments of pBR322 DNA Containing the *Bam* HI Site

Fragment Length base pairs	Restriction Digests	Distance of the *Bam* HI Cleavage Site from Termini
74	*Taq* I - *Hae* II	36, 38
376	*Sph* I - *Eco* RV	190, 186
650	*Eco* RI - *Sal* I	375, 275
870	*Dde* I - *Ava* II	447, 423
1000	*Dde* I - *Bgl* I	447, 553
1500[a]		740, 760
2278	*Ava* I - *Pst* I	1129, 1049
4362	*Nde* I	2441, 1921

[a] This fragment was produced by the digestion of the 2278 base pair fragment with *Bal*-31 nuclease.

The kinetic behavior of the methylase differed from the endonuclease under similar reaction conditions. At DNA concentrations of 1 nM in the absence of NaCl the velocity of methylation decreased in the range of 74 to 1000 base pairs but increased from 1500 to approximately 4300 base pairs (Figure 12A). In the presence of 200 mM NaCl this velocity increase was eliminated (Figure 12B). It is not yet clear why this behavior occurs. The binding of AdoMet or the overall equilibrium affinity for DNA may vary with substrate length. The increase in reaction rates with longer fragments could have been a result of facilitated diffusion since they were sensitive to high NaCl concentrations. The trend of decreasing methylation rates between 74 and 1000 base pairs was reversed by higher DNA concentrations (12 nM) but only a slight increase in velocity was observed with substrates longer than 74 base pairs (Figure 12C). At these DNA concentrations increases in the NaCl concentration to 120 mM caused a significant increase in the long chain preference of the methylase (Figure 12C). This phenomenon was also seen in reactions containing equimolar amounts of long and short substrates (Figure 12D). An increasing kinetic preference for the longer DNA up to 120 mM NaCl was observed and was still apparent at 200 mM NaCl.

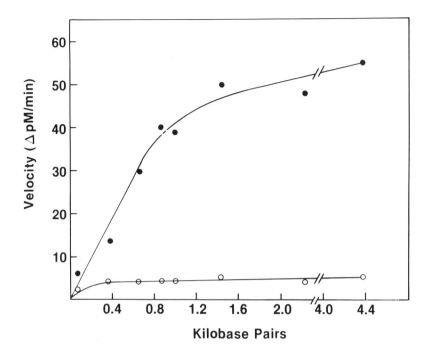

Figure 11: Kinetic Behavior of *Bam* HI Endonuclease with DNA Substrates of Different Lengths.

The initial velocity of *Bam* HI endonuclease was examined in reactions containing 1 nM DNA fragments in the range of 74 to 4362 base pairs. The reactions were run under standard conditions (●) or in the presence of 150 mM NaCl (○).

By providing competitive binding sites non-specific flanking DNA can slow down specific association. Alternatively, specific association can be accelerated by facilitated diffusion via non-specific DNA. Theoretical results predict a range of non-specific binding constants that promote optimal increases in specific association rate constants [31]. Predictions derived from a sliding model have been verified by experiments with *lac* repressor which demonstrated that increases in KCl concentrations to 100 mM promoted increases in the specific association rate constants with long DNA to maximum values [29]. Increases in KCl past this value drastically lowered association rate constants below their original values in the low salt case. Since salt strongly affects the non-specific binding constant these results suggest that a marginal decrease in non-specific binding affinity allows for a greater amount of productive interactions that lead to sliding as opposed to the protein being trapped in an unproductive, non-specific binding mode. A similar situation might be occurring with *Bam*H I methylase and could explain its enhanced long chain preference at NaCl concentrations of 100 to 120 mM. Figure 12D indicates residual long chain preference at NaCl concentrations as high as 200 mM. This comparatively refractory behavior to high salt concentrations at high DNA concentrations is consistent with the persistence of the differences between methylation rates with the EcoR I and Nde I linearized substrates (Figure 10F). At high DNA concentrations residual non-specific binding leading to facilitated transfer may still occur at high salt concentrations.

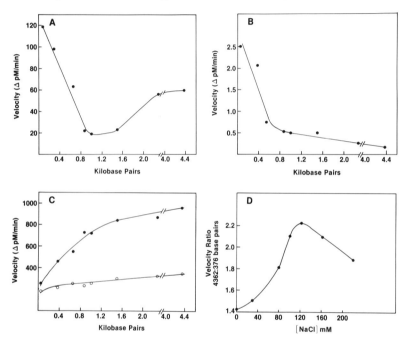

Figure 12: <u>Kinetic Behavior of Bam HI Methylase with DNA Substrates of Different Lengths.</u>

The initial velocity of Bam HI methylase was examined with pBR322 DNA fragments in the range of 74 to 4362 base pairs. The AdoMet concentration was 35 uM.
    A.   The concentrations of DNA and methylase were 1 and 0.15 nM respectively. The reactions contained no added NaCl.
    B.   As in A except the NaCl concentration was 200 mM.
    C.   The concentrations of DNA and methylase were 12 and 0.5 nM respectively. Reactions had no added NaCl (O) or 150 mM NaCl (●).
    D.   Ratios of the initial velocity of methylation with the 4362:376 base pair substrates at increasing NaCl concentrations. Each reaction contained both fragments at a concentration of 12 nM and an enzyme concentration of 0.5 nM.

The kinetic behavior of the endonuclease and the methylase suggests that their non-specific interactions with DNA are different. This idea was addressed with inhibition studies using the 74 base pair substrate and a purified fragment of pBR322 DNA lacking the BamH I site. In the range of 0.1 to 1 uM non-specific base pairs only a slight inhibition of methylase activity was observed (Figure 13). In contrast, the endonuclease activity was inhibited 20% in the same concentration range. In the range of 2 to 75 uM non-specific base pairs the rapid increase in methylase inhibition was considerably greater than that of the endonuclease. The endonuclease exhibited a biphasic pattern of inhibition which may indicate that the enzyme possesses two binding sites with different affinities for non-specific DNA. Saturation of the stronger site results in 20% inhibition of activity while saturation of the weaker site almost totally inhibits activity. One of these sites may be the active site and both sites would be interactive assuming this interpretation is valid. The methylase might bind two non-specific DNA fragments cooperatively as judged by the abrupt increase in inhibition at 2 uM base pairs. These competition curves indicate that the enzymes' non-specific association with DNA are qualitatively different. This may account for some of the differences seen in the previous kinetic experiments. However, on the basis of initial velocity kinetics both enzymes share approximately the same binding affinity for the BamH I site. The difference in Kms of the enzymes for the EcoR I and Nde I linearized pBR322 DNA substrates is supportive evidence for the occurrence of facilitated transfer since this process ultimately affects specific binding. However, the difference in Km values is too large by a factor of 2.6 (based on theoretical calculations [31]) if a sliding mechanism was solely responsible for the differences in reaction rates. The DNA conformation in the vicinity of the BamH I site and other facilitated diffusion processes in addition to sliding may play a role.

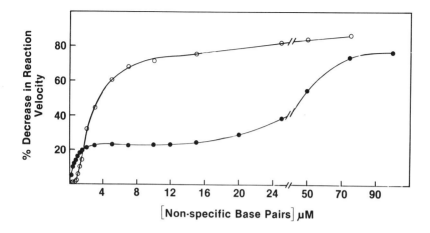

Figure 13: Inhibition of *Bam* HI Endonuclease and Methylase Activity with Non-specific DNA.

The initial velocity of cleavage (●) and methylation (O) was determined in reactions containing the 74 base pair substrate at 1 nM and various concentrations (expressed in terms of base pairs) of a 3986 base pair fragment of pBR322 DNA not containing the *Bam* HI site. The percent decrease in velocity was calculated from reactions run in the absence of non-specific DNA.

IV. AMINO ACID RESIDUES REQUIRED FOR ACTIVITY

A. *Bam*H I Endonuclease

*Bam*H I endonuclease was found to have sensitivity towards reagents that modify sulfhydryl groups. At neutral pH and 37°C incubation of the enzyme with 6 mM iodoacetamide, dithiobis(2-nitro-benzoic acid) or N-ethylmaleimide for 45 minutes resulted in 45%, 23% and 30% inhibition of activity respectively. Enzyme activity was also inhibited 60% by a 45 min. incubation with 6 mM carbodiimide, suggesting that aspartate and glutamate residues are important (although the possibility of side reactions with cysteine and tyrosine has not been ruled out).

Arginine residues have been reported to be important in enzymes containing nucleotide and phosphate binding sites [34-38]. We have investigated the importance of arginine in *Bam*H I endonuclease using the reagent butanedione [17] according to the method of Riordan [39]. Enzyme activity was found to be progressively inhibited by increasing concentrations of butanedione in the presence of 50 mM sodium borate, pH 8.0. There was a concomitant loss of DNA binding ability as judged by nitrocellulose filter assays. The inhibition showed an absolute dependence on borate and could be reversed to the extent of 85% by dialysis of modified enzyme against borate-free buffers. These results indicate that the butanedione was not causing significant non-specific denaturation of the enzyme.

Preincubation of the endonuclease with low concentrations of pBR322 DNA (one BamH I site) or Col EI DNA (no BamH I site) resulted in efficient protection against the butanedione modification (Figure 14). Non-specific binding of the enzyme was probably responsible for the protection generated by the Col EI DNA since preincubation with this DNA in the presence of 150 mM NaCl reduced protection to levels 50% lower than those generated by pBR322 DNA under identical conditions. Negatively charged polymers such as heparin and RNA competitively inhibit BamH I endonuclease[18] implicating the involvement of electrostatic interactions during non-specific binding at the active site. Considering that specific binding could be 5 to 6 orders of magnitude stronger than non-specific binding (as in the case of EcoR I endonuclease [40]) it seems odd that Col EI DNA protected the enzyme as well as pBR322 DNA (on the basis of nucleotide concentration). In these experiments non-specific nucleotide concentrations were in the range of 5×10^{-8} to 6×10^{-7} M while endonuclease concentrations were approximately 2×10^{-7} M (dimer). Only a small fraction of the enzyme would be specifically bound with pBR322 DNA. Specifically bound and, presumably, specifically protected enzyme would be difficult to distinguish since far more enzyme could be non-specifically associated with near equimolar amounts of non-specific sites. This argument assumes that the kinetics of the butanedione reaction with free enzyme is slow and inefficient compared to non-specific binding between enzyme and DNA.

As previously mentioned the deoxydinucleotide subsets of the BamH I sequence are specific inhibitors and were tested for their ability to protect the endonuclease from the butanedione modification. The dinucleotides GG, GA and AT generated various degrees of protection (Figure 15). Protection conferred by GG was substantially more efficient than GA and AT. No protection against inactivation was given by TC and CC. Unrelated subsets such as TA and GC also failed to protect the enzyme. Dinucleotides corresponding to a 5' flanking base and the first guanine of the BamH I sequence (CG and AG) gave the second highest amount of protection produced by the inhibitors. Non-linear least squares analysis of dinucleotide concentration versus percent protection yielded hyperbolic curves. The concentrations of GG, CG, AG, GA and AT that resulted in 50% protection were 0.3, 9.1, 12.7, 64.7 and 159.2 uM respectively.

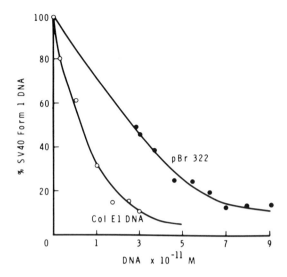

Figure 14: Effect of pBR322 and Col EI DNAs on the Butanedione Modification of Bam HI Endonuclease.

Bam HI endonuclease was incubated with various concentrations of pBR322 (●) or Col EI (○) DNAs for 15 min. at 4°C. The enzyme was then exposed to 30 mM butanedione and 50 mM borate, pH 8.0, for 30 min. in the dark at 22°C. Aliquots were removed and enzyme activity was determined with form I ^{3}H-SV40 DNA. Decreasing form I content indicates increasing protection against butanedione deactivation.

The protective effects of DNA and the loss of DNA binding ability after the butanedione modification suggested that some of the modified arginine residues may be located in the enzyme's active site. In addition, the following hierarchy of dinucleotide protective ability was found: GG>CG>AG>GA>AT>TC≥CC. The superior protective ability of the competitive inhibitor GG also suggests that active site arginine residues may participate in the recognition of the guanines of the BamH I palindrome. Furthermore, the relatively efficient protective ability of CG and AG suggests that interactions between arginine and the first guanine of the sequence may be of particular importance. Previous studies have shown that AG is an inhibitor of BamH I endonuclease with a K_I of 114 uM [18]. The low amounts of protection provided by GA and AT may be due to proximity effects that sterically hinder access of butanedione or to changes in local chemical reactivities during dinucleotide binding. The dinucleotide CC was the most potent inhibitor of the endonuclease (Table III) but failed to generate any protection. This is consistent with a DNA binding site composed of several kinds of amino acid functionalities and binding domains designed to interact with different regions of the BamH I sequence. However, these studies cannot eliminate the possibility of substrate or inhibitor induced conformational changes that affect the reactivity of arginine residues which are located outside the active site but are important to the structural integrity of the endonuclease.

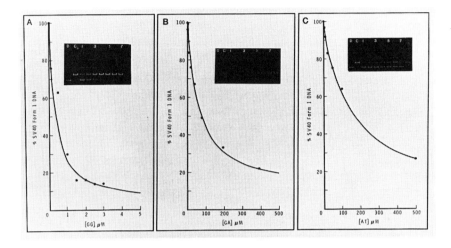

Figure 15: Protection of Bam HI Endonuclease against Butanedione Deactivation by the 5'-phosphoryl deoxydinucleotide Inhibitors GG, GA and AT.

Bam HI endonuclease was incubated with various concentrations of the different dinucleotides for 15 min. at 4°C. and subsequently exposed to butanedione as described. Aliquots were removed and enzyme activity was determined. Decreasing form I DNA content indicates increasing protection against butanedione deactivation. Insets represent cleavage products after incubation with 0.25 to 3 uM GG (A), 5 to 500 uM GA (B) and AT (C) and modification with butanedione.

Sequence specific interactions between proteins and DNA have been theoretically explained in terms of putative hydrogen bond contacts between amino acid residues and the major groove moieties of the bases[41]. It has been suggested that the arginine residues can form stereochemically favorable hydrogen bond complexes with the O^6 and N^7 acceptors of guanine [42]. Theoretical calculations show that arginine forms more stable hydrogen bond complexes with single stranded guanines and the guanine of G·C base pairs than glutamate, aspartate or their amides [43]. It is also possible that the arginines participate in electrostatic interactions with the phosphates of the guanine nucleotides. Evidence for such electrostatic contacts has been obtained with EcoR I endonuclease [13].

B. *Bam*H I Methylase

*Bam*H I methylase was found to be very sensitive to sulfhydryl modifying reagents. Complete inhibition of activity occurred within 20 min., at 4°C, after incubation with 1.5 mM N-ethylmaleimide or 30 uM p-chloromercuribenzoate. The activity inhibited by p-chloromercuribenzoate could be completely recovered after a 30 sec. incubation with 1 mM 2-mercaptoethanol. Preincubation with AdoMet or AdoHcy protected the methylase from modification by both reagents. The pattern of protection generated by increasing concentrations of AdoMet or AdoHcy was complex, having 3 maxima occurring at 4, 12 and 38 uM. AdoMet protected the enzyme more efficiently than AdoHcy. DNA gave minimal protection (10%) but this low level of protection increased in an apparently hyperbolic manner in the range of 0.1 to 25 nM DNA. pUC9DNA (one *Bam*H I site), *Bam*H I methylated pUC9DNA and Col EI DNA (no *Bam*H I site) all generated approximately 10% protection at concentrations of 25 nM. Interestingly, DNA was antagonistic to the protection conferred by AdoMet or AdoHcy. For instance, incubation with 35 uM AdoHcy and 10 nM pUC9DNA lowered protection by 40% compared to the AdoHcy alone. Similarly, incubation with 35 uM AdoMet and 10 nM Col EI DNA lowered protection by 50%. Conformational changes in *Bam*H I methylase upon substrate binding may affect the reactivity of various cysteine residues and this may be responsible for the unusual protection profiles observed.

*Bam*H I methylase also exhibited a borate dependent sensitivity to butanedione (Figure 16A). Substrate protection studies revealed striking similarities to *Bam*H I endonuclease. Both specific and non-specific DNA efficiently protected the methylase from the butanedione modification (Figure 16B) and the deoxydinucleotide inhibitors GG (Figure 16C), and GA and AT (Figure 16D) also conferred protection. As in the case of the endonuclease GG protected the methylase most efficiently. TC and CC conferred no protection and the unrelated subsets TT, AA and GC were also ineffective. The presence of AdoMet or AdoHcy did not generate any protection and appeared to have no effect on the protective ability of DNA or the dinucleotides. These studies suggest that arginine residues in both the endonuclease and the methylase may be important to the specific interactions with the *Bam*H I recognition site.

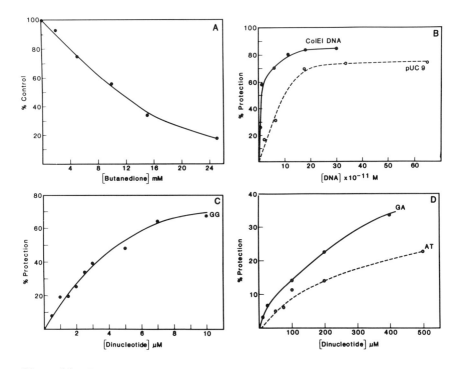

Figure 16: **Butanedione Modification of _Bam_ HI Methylase**.

In all experiments aliquots of the incubation mixtures containing modified enzyme were removed and assayed in reactions containing 20 nM pUC9 DNA and 10 uM ^{3}H-AdoMet.

- A. The methylase was incubated with various concentrations of butanedione in 50 mM borate, pH 8.0, in the dark for 20 min. at 22°C. and activity was determined as described.

- B. The methylase was incubated with various concentrations of pUC9 DNA (O) or Col EI DNA (●) at 4°C. The enzyme was then exposed to 20 mM butanedione and 50 mM borate, pH 8.0, for 20 min at 22°C. and the amount of protection against enzyme deactivation was determined.

- C. Protection of methylase against the butanedione modification (as described in B) by the 5'-phosphoryl deoxydinucleotide GG.

- D. Protection of methylase against the butanedione modification by the 5'-phosphoryl deoxydinucleotides GA and AT.

V. PERTURBATION OF SEQUENCE SPECIFICITY BY ORGANIC SOLVENTS

A. BamH I Endonuclease

Highly purified preparations of BamH I endonulcease were found to have a trace amount of a second, distinct cleavage activity. Separation of these activities by fractionation with a variety of approaches failed. Further experimental work provided strong evidence that both the normal BamH I activity and the second activity, termed BamH I.1, were generated by the same protein [44]. A similar phenomenon was also reported by Polisky, et al, with EcoR I endonuclease [45]. The secondary activity called EcoR I* could be induced by alterations in the ionic environment and pH. Similar changes in specificity have been noted for Bst I [46] and Bsu I [47] endonucleases.

In the absence of monovalent cations increases of the reaction pH to 9.0 enhanced BamH I.1 activity. The addition of glycerol or dimethylsulfoxide to the reaction caused the greatest enhancement of BamH I.1 activity (Figure 17). Both solvents caused the enzyme to generate identical DNA fragmentation patterns. Ethylene glycol, ethanol and dioxane produced the same effects. BamH I endonuclease normally cleaves lambda DNA five times. In the presence of 36% glycerol or 17% dimethylsulfoxide the amount of fragments produced was greater than 20. ϕX174RF DNA is not normally cleaved by BamH I endonuclease but is cleaved twice by the second activity. The addition of NaCl greatly suppressed BamH I.1 activity. High enzyme to DNA ratios and alkaline pHs in the presence of organic solvents served to optimize the secondary activity. The conversion of BamH I to BamH I.1 was found to be reversible.

The physical relationship between BamH I and BamH I.1 was tested by several approaches. The ratio of BamH I to BamH I.1 activity was determined at a number of different glycerol concentrations and revealed a parent-daughter relationship as described by the equation $A_1 = A_0 (1 - e^{-bG})$ where b = slope of BamH I activity and G = fraction of glycerol in the reaction. The logarithm of both activities plotted against the glycerol concentration generated lines of equal slopes, implying that both activities are catalyzed by the same protein. Modification of lysine residues with pyridoxal phosphate or arginine residues with butanedione equally inhibited both activities. Thermal denaturation studies performed between 37 and 63°C showed a concomitant loss of both activities.

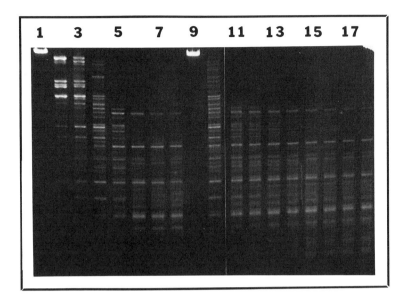

Figure 17: <u>Effect of Glycerol and Dimethylsulfoxide on Bam HI Endonuclease</u>.

Glycerol (lanes 1 to 8) or dimethylsulfoxide (Lanes 10 to 18) were introduced into the Bam HI reaction mixture containing lambda DNA. Lanes 1 and 9 are undigested lambda DNA. Lanes 2 to 8 were reaction mixtures containing 1 to 61% (v/v) of glycerol in increments of 10% that were incubated for 1 h. at 37 deg. C. Lanes 10 to 18 were reaction mixtures containing 17% (v/v) dimethylsulfoxide with incubation times of 20 to 180 min. in increments of 20 min. Cleavage products were fractionated on 1% agarose gels.

The major BamH I.1 sites on pBR322 and SV40 DNAs were mapped and sequenced [48]. These studies showed that the sites were subsets of the BamH I sequence and were primarily of the type 5'-GGANCC-3', 5'-GGNTCC-3' and 5'-GNATCC-3'. The relative rates of catalysis for strong BamH I.1 sites was determined (Figure 18). The order of the rates of digestion was GGATCC>GGGTCC>GGCTCC>GGTTCC.

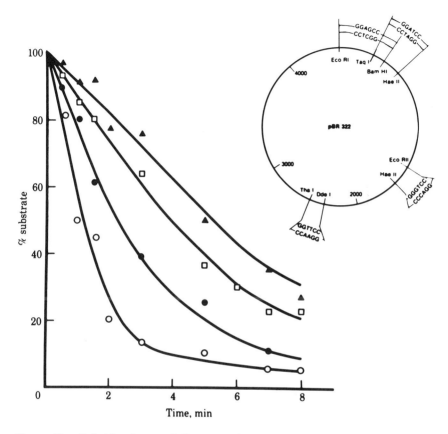

Figure 18: <u>Relative Rates of Cleavage of the Strong *Bam* HI.1 Sites by Bam HI Endonuclease</u>.

Restriction fragments from pBR322 DNA containing a single *Bam* HI.1 site were end-labeled with ^{32}P. Reactions containing equimolar quantities of DNA were run in the presence of 36% glycerol. Aliquots were removed at various time intervals and analyzed by polyacrylamide gel electrophoresis. The substrates are represented by GGATCC (○), GGGTCC (●), GGCTCC (□) and GGTTCC (▲).

Seeman, et. al.[41] proposed that the information required for specific binding of a protein to a nucleic acid residues in the pattern of hydrogen bond donors and acceptors possessed by the individual base pairs. They proposed that a protein must make at least two hydrogen bond contacts per base pair if there is to be unambiguous discrimination. Woodbury, et. al., developed a schematic illustration of these kinds of base pair interactions [49]. Such a representation of putative hydrogen bond contacts between BamH I endonuclease and the A·T base pair in its recognition site is depicted in Figure 19A. The relative positions available for hydrogen bond interactions of the A·T base pair in the BamH I sequence and a T·A base pair are identical in the minor groove but are inverted in the major groove (Figure 19A). BamH I and BamH I.1 do not cleave the sequence GGTACC suggesting that BamH I discrimination of A·T and T·A base pairs occurs in the major groove. Berkner and Folk have shown that uracil substituted DNA does not significantly affect the kinetics of BamH I endonuclease [50] implying that the thymidine methyl group is not essential. Similarly, the methylation of the 6-amino groups of the adenines does not adversely affect BamH I cleavage. The N^7 of adenine and the O^4 of thymidine are the only remaining positions in the major groove available for hydrogen bonding at this central position in the sequence.

The rate variations among the different BamH I.1 sites (Figure 18) may be due to differences in hydrogen bond interactions at the A·T region of the major groove. Relaxation of specificity may involve fewer hydrogen bond contacts between the sequence and the enzyme. With the exception of the canonical BamH I sequence, the BamH I.1 sequence that is cleaved fastest is GGGTCC. This site maintains a N^7 purine contact point in the same relative position. Consequently, three of the four contacts (N^7 of adenine, O^4 of thymidine and the N^7 of the guanine) are preserved and the single contact with the O^4 of thymidine is lost by the cytosine substitution. In the case of GGAGCC and GGAACC two of the four original contacts are lost, the N^7 of adenine and the O^4 of thymidine (Figure 19B). These sequences have the lowest cleavage rates among the sites examined. The observed rate differences between these latter two sites may be due to the substitution of the thymidine O^4 by the guanine O^6 in GGAGCC. This substitution maintains a hydrogen bond acceptor which may be in a stereochemically similar position and could explain the kinetic preference of GGAGCC over GGAACC.

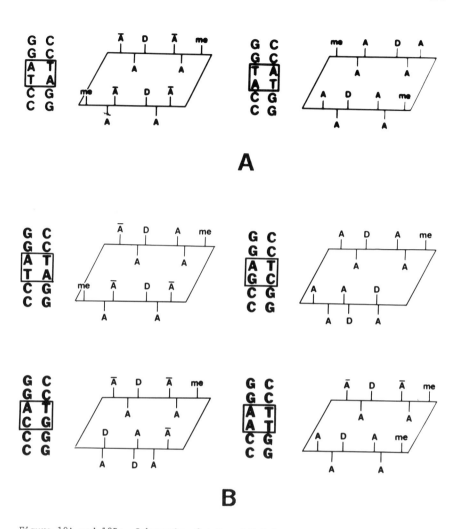

Figure 19A and 19B: <u>Schematic of A T and T A Base Pairs.</u>

The schematic illustrates methyl groups (me) and positions available for hydrogen bond contacts between the DNA and the enzyme. A and D represent hydrogen bond acceptors and donors respectively. Overbars indicate proposed sites for Bam HI.1 interactions. Acceptors are the N^7 of adenine and guanine, O^4 of thymidine and O^6 of guanine. Donor positions are the 6-amino of adenine and 4-amino of cytosine. The horizontal line represents the hydrogen bond between the N^1 of the pyrimidine and the N^3 of the purine base pair. Acceptors and donors above the horizontal line face into the major groove of the DNA and those below are in the minor groove.

B. *Bam*H I Methylase

Comparitive studies performed with *Bam*H I methylase demonstrated a glycerol induced perturbation of sequence specificity [7]. In the presence of 30% glycerol hypermethylation of lambda and SV40 DNAs was observed. ϕX174RF DNA, which has no *Bam*H I site, was found to incorporate approximately 3 moles of methyl groups per mole of DNA in the presence of glycerol. Hypermethylation was favored by alkaline conditions and high enzyme to DNA ratios. In contrast to the endonuclease, secondary methylating activity was optimized by 100 mM KCl or NaCl. Time course experiments performed with 30% glycerol showed only a limited amount of methyl group incorporation with pBR322 DNA that had been previously cleaved with *Bam*H I.1 endonuclease activity (Figure 20). This suggests that the majority of *Bam*H I.1 sites acted on by the endonuclease are also methylated in the presence of glycerol. The low amount of methyl group incorporation that did occur was probably due to incomplete *Bam*H I.1 cleavage although secondary sites unique to the methylase may also exist. The specificity of *Eco*R I methylase is also disturbed by glycerol and the enzyme appears to share the same EcoR I* recognition sequences with the endonuclease [51].

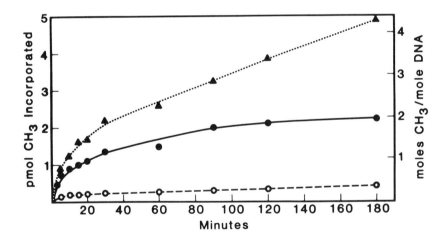

Figure 20: *Bam* HI Hypermethylation in the presence of Glycerol.

Bam HI methylase was added to reactions containing 25 mM Tris-HCl, pH 8.5, 10 mM 2-mercaptoethanol, 100 mM KCl, 200 ug/ml BSA, 3 uM ^{3}H-AdoMet with form I pBR322 DNA (●), form I pBR322 DNA in the presence of 30% glycerol (▲), *Bam* HI.1 cleaved pBR322 DNA in the presence of 30% glycerol (O). The DNA concentration was 6nM.

The pH optimum for BamH I hypermethylation and BamH I.1 cleavage is in the vicinity of pH 8.5 to 9.0. However, unlike the case of the endonuclease, this pH range is not the optimum for the methylase under normal conditions. This suggests that DNA binding and recognition parameters of the methylase may differ for BamH I.1 sites. Also, conformational and ionization changes in the protein may be required for efficient hypermethylation. BamH I methylase best catalyzed hypermethylation under conditions of comparatively high ionic strength. The endonuclease was most sensitive to the effects of glycerol (and other organic solvents) at low ionic strength. It is not clear why low ionic strength promotes aberrant cleavage. BamH I.1 cleavage is not kinetically favored over that which occurs at the canonical site, suggesting relatively unfavorable binding energetics. Minimization of the important electrostatic component of binding by high salt concentrations could further exacerbate an already unfavorable binding problem at non-canonical sites. It seems reasonable to assume that electrostatic interactions are important to methylase:DNA in interactions as well. Yet the high salt requirement for BamH I hypermethylation appears to be contrary to presumably less favorable binding at BamH I.1 sites. Preliminary kinetic studies have shown that the Vmax of BamH I methylase increases with increasing salt concentrations in the range of 0 to 100 mM. Thus, the enzyme may need salt to facilitate conformational and catalytic prerequisites for optimum activity regardless of the presence of glycerol. Alternatively, methylase binding to secondary sites may not depend as heavily on electrostatic interactions as the endonuclease does.

Glycerol induced relaxation of sequence specificity apparently affects the BamH I and EcoR I enzymes differently. Recognition of the internal base pairs of the BamH I sequence is altered while the base wobbles occur at the external portions of the EcoR I sequence. The external bases of the BamH I sequence may therefore be the more critical elements to BamH I endonuclease and methylase site discrimination. Consistent with this idea are the low K_{is} found with the dinucleotide inhibitors GG and CC as compared to the other subsets. EcoR I* sites are methylated in vivo as judged by the low level of methyl group incorporation catalyzed by EcoR I methylase in the presence of glycerol with E. coli RY13 (r^+m^+) chromosomal DNA [51]. In contrast, chromosomal DNA from B. amyloliquefaciens H is substantially cleaved by BamH I.1 activity. Since BamH I hypermethylated DNA is largely resistant to cleavage by BamH I.1 activity in vitro, the implication is that in vivo hypermethylation in this system does not occur to a significant extent.

REFERENCES

1. Wilson, G.A. and Young, F.E. (1977) J. Mol. Biol. 97, 123.
2. Roberts, R.J., Wilson, G.A. and Young, F.E. (1977) Nature (London) 265, 85.
3. Hattman, S., Keister, T. and Gottehrer, A. (1978). J. Mol. Biol. 124, 701.
4. Mann, M.B. and Smith, H.O. (1977) Nucl. Acids. Res. 4, 4211.
5. Pirotta, V. (1976) Nucl. Acids Res. 3, 1747.
6. Smith, L.A. and Chirikjian, J.G. (1979) J. Biol. Chem. 254, 1003.
7. Nardone, G., George, J. and Chirikjian, J.G. (1984) J. Biol. Chem. 259, 10357.
8. Sussenbach, J.S., Monfoort, C., Schipnof, R. and Stubberingh, E.E. (1976) Nucl. Acids Res. 3, 3193.
9. Stubberingh, E.E., Schipnof, R. and Sussenbach, J.S. (1977) J. Bact. 131, 645.
10. Lui, A.C.P., McBride, B.C., Vovis, G.F. and Smith, M. (1979) Nucl. Acids Res. 6, 1.
11. Brooks, J. and Roberts, R.J. (1982) Nucl. Acids Res. 10, 913.
12. Modrich, P. and Zabel, D. (1976) J. Biol. Chem. 251, 5866.
13. Jen-Jacobsen, L., Kurpiewski, M. Lesser, D., Grable, J., Boyer, H.W., Rosenberg, J.M. and Greene, P.J. (1983) J. Biol. Chem. 258, 14638.
14. Chou, P.Y. and Fasman, G.D. (1978) Adv. Enzymol. 47, 45.
15. Rose, G.D. (1978) Nature 272, 586.
16. Lee, Y. and Chirikjian, J.G. (1979) Fed. Proc. 38, 294.
17. George, J., Nardone, G. and Chirikjian, J.G. (1985) J. Biol. Chem.
18. Hinsch, B., Mayer, H. and Kula, M.R. (1980) Nucl. Acids Res. 8, 2547.
19. A-Lien, L., Jack, W.E. and Modrich, P. (1981) J. Biol. Chem. 256, 13200.
20. Lee, Y. and Chirikjian, J.G., unpublished observations.
21. Berman, M., Shahn, E., Weiss, M.F. (1962) Biophys. J. 2, 275.
22. Rubin, R.A. and Modrich, P. (1978) Nucl. Acids Res. 5, 2991.
23. Baumstark, B.R., Roberts, R.J. and RajBhandary, U.L. (1979) J. Biol. Chem. 254, 8943.
24. Gunthert, U., Jentsch, S. and Freund, M. (1981) J. Biol. Chem. 256, 9346.
25. Yuan, R. (1981) in Gene Amplification and Analysis, Volume I, (Chirikjian, J.G., ed.) Elsevier/North Holland, N.Y., pp. 64-66.
26. Walsh, C. (1979) Enzymatic Reaction Mechanisms, W.H. Freeman and Co., San Francisco, p. 851-857.

27. Campos, G., Guixe, V. and Babul, J. (1984) J. Biol. Chem. 259, 6147.
28. Segal, I. (1975) Enzyme Kinetics, Wiley Interscience, N.Y., pp. 623-625.
29. Winter, R.B., Berg, O.G. and von Hippel, P.H. (1981) Biochemistry 20, 6961.
30. Jack, W.E., Terry, B.J. and Modrich, P. (1982) Proc. Natl. Acad. Sci., U.S.A. 79, 4010.
31. Berg, O.G., Winter, R.B. and von Hippel, P.H. (1981) Biochemistry 20, 6929.
32. Nardone, G., George, J. and Chirikjian, J.G. (1986) J. Biol. Chem., in press.
33. Terry, B.J., Jack, W.E. and Modrich, P. (1985) J. Biol. Chem. 260, 13130.
34. Marcus, F., Schuster, S.M. and Lardy, H.A. (1976) J. Biol. Chem. 251, 1775.
35. Frigeri, L., Galante, Y.M. and Hatefi, Y. (1978) J. Biol. Chem. 253, 8935.
36. Salvo, R.A., Serio, G.F., Evans, J.E. and Kimball, A.P. (1976) Biochemistry 15, 493.
37. Armstrong, V.W., Sternback, H. and Eckstein, F. (1976) FEBS Lett. 70, 48.
38. Borders, L.L., Riordan, J.F. and Auld, D.S. (1975) Biochem. Biophys. Res. Commun. 66, 490.
39. Riordan, J.F. (1973) Biochemistry 12, 3915.
40. Terry, B.J., Jack, W.E., Rubin, R.A. and Modrich, P. (1983) J. Biol. Chem. 258, 9820.
41. Seeman, N.C., Rosenberg, J. and Rich, A. (1976) Proc. Natl. Acad. Sci., U.S.A. 73, 804.
42. Helene, C. (1977) FEBS Lett. 74, 10.
43. Kumar, N.V., and Govil, G. (1984) Biopolymers 23, 1995.
44. George, J., Blakesley, R.W. and Chirikjian, J.G. (1980) J. Biol. Chem. 255, 6521.
45. Polisky, B., Greene, P., Garfin, D.E., McCarthy, B.J., Goodman, H.M. and Boyer, H.W. (1975) Proc. Natl. Acad. Sci. U.S.A. 72, 3310.
46. Clarke, C.M. and Hartley, B.S. (1979) Biochem. J. 177, 49.
47. Heininger, K., Hort, W. and Zachau, H.G. (1977) Gene (Amst.) 1, 291.
48. George, J. and Chirikjian, J.G. (1982) Proc. Natl. Acad. Sci. U.S.A. 79, 2432.
49. Woodbury, C.P., Hagenbuchle, O. and von Hippel, P.H. (1980) J. Biol. Chem. 255, 11534.

50. Berkner, K.L. and Folk, W.R. (1979) J. Biol. Chem. 254, 2551.
51. Woodbury, C.P., Downey, R.L. and von Hippel, P.H. (1980) J. Biol. Chem. 255, 11526.

6

The *Eco*R V Restriction Endonuclease

Peter A. Luke*, Sarah A. McCallum**, and Stephen E. Halford**

*Anglian Biotechnology Ltd.
Hawkins Road, Colchester COZ 8JX (U.K.)

**Department of Biochemistry
Unit of Molecular Genetics
University of Bristol
Bristol BS8 1TD (U.K.)

ABSTRACT

Type II restriction endonucleases have attracted attention for two main reasons: firstly, their many applications in the dissection of DNA and in the construction of novel DNA molecules; secondly, as systems for studying the interactions of proteins with specific DNA sequences. With respect to the latter, the EcoR I restriction endonuclease has been examined in greater depth than any other type II enzyme [1-3]. However, the EcoR I enzyme has a major disadvantage as a system for studying DNA-protein interactions: the protein has a remarkably low solubility. The solutions in which EcoR I shows maximal activity, and also affinity for its recognition site, are saturated at less than 0.5 μM of this protein [4]. Consequently, many techniques that have been developed to study protein-ligand interactions but which require high concentrations of the protein in solution, such as NMR spectroscopy, cannot be used on EcoR I. But this drawback does not apply to all type II restriction enzymes. A different enzyme, the EcoR V restriction endonuclease [5-7], has special advantages as a system for studying DNA-protein interactions. In particular, this is the only type II restriction enzyme (apart from EcoR I [3]) for which crystals of the protein have been reported [7].

II. INTRODUCTION: DNA CLEAVAGE SITE

Kholmina et al [5] found a novel restriction endonuclease in a strain of Escherichia coli, J62[pLG74], that have been derived by the conjugation of R plasmids from a clinical isolate of E. coli. This enzyme, named EcoR V, was shown to cleave DNA at the sequence

5' - G A T↓A T C - 3'
3' - C T A↑T A G - 5'

Kholmina et al [5] had proposed that EcoR V cleaves the phosphodiester bonds between the fifth and sixth bases from the 5' end of each strand of the DNA, thus leaving the DNA with 3' tetranucleotide extensions. But subsequent studies have demonstrated that EcoR V cleaves its recognition sequence at the sites marked above by arrows, and thus leaves the DNA with blunt ends [6,7]. This was shown by both DNA sequence analysis of a derivative of phage M13 mp8 that had been cleaved by EcoR V [6], and by chemical analysis of the products that EcoR V generated from the synthetic oligonucleotide, CCGATATCGG [7].

No other restriction enzyme, apart from one in a different strain of E. coli, recognizes the same DNA sequence as EcoR V [8]. Consequently, this enzyme is useful for the analysis of DNA by restriction mapping and related techniques. The EcoR V enzyme also has applications in the construction of recombinant DNA. The plasmid pBR322, one of the most widely used vectors for recombinant DNA, confers resistance to both ampicillin and tetracycline but has a single recognition site for EcoR V within the gene for tetracycline resistance [8]. Hence, the cloning of DNA at the EcoR V site on pBR322 may be monitored by the inactivation of tetracycline resistance. The same applies to a number of other vectors derived from pBR322, such as pBR328 or pAT153.

II. MOLECULAR GENETICS OF THE EcoR V RESTRICTION/MODIFICATION SYSTEM

The genes that specify the EcoR V R/M system have been identified and their structures were determined by Bougueleret et al [9]. The strain that produced the EcoR V restriction enzyme, E. coli J62[pGL74], contained several different plasmids. One of these, a 6.2 kb plasmid now named pLB1, was found by transformation experiments to carry the R/M system: transformants were selected by their ability to restrict phage λ. Restriction fragments from pLB1 were subsequently cloned in pBR322 and the smallest of these that was found to carry the complete R/M system, pLB83, contained a 3.0 kb Pvu II - Bgl II fragment from pLB1 (Figure 1). The positions of the genes for the EcoR V restriction endonuclease and modification methylase were located by using nuclease Bal 31 to generate deletions from the Pvu II end of this fragment. Deletion of the first 880 bp affected neither activity (pLB6: Figure 1) but all deletions within the next 1100 bp abolished EcoR V restriction activity without affecting the methylase (pLB7: Figure 1). All further deletions removed both activities [9].

The region of DNA that codes for the EcoR V R/M system was further characterized by determining its complete DNA sequence [9]. This revealed only two significant open-reading frames, one of 735 bp at the locus assigned to the endonuclease gene, and a second of 986 bp at the position of the gene for the methylase. The two open reading frames are on opposite strands of the DNA and are separated by an intergenic region of 310 bp. The 735 bp reading frame was confirmed as that for the EcoR V nuclease by determining the amino acid sequences for both the first 40 N-terminal residues and the last 11 C-terminal residues of the protein: apart from the absence of the N-terminal methionine (presumably removed by post-synthetic processing), the amino acid sequences were as predicted from the DNA sequence [9]. A similar analysis confirmed that the 896 bp reading frame coded for the EcoR V methylase [9]. As with other type II R/M systems [10,11], neither gene nor protein sequence for the nuclease showed any homology with those for the methylase.

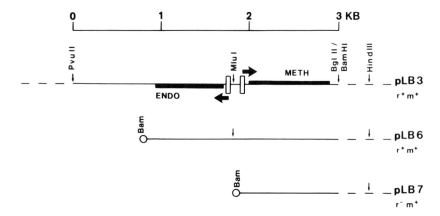

Figure. 1. The genes for the EcoR V R/M system [9]. The naturally occurring plasmid that codes for the EcoR V M/R system, pLB1 (not shown), yielded a 3.0 kb Pvu II - Bgl II restriction fragment. On the recombinant plasmid, pLB3, part of which is shown above, this fragment (solid line) is carried between the Pvu II and BamH I sites of pBR322 (dashed line). On pLB3, the position of the genes for the EcoR V restriction endonuclease (ENDO) and modification methylase (METH) are marked by thicker lines and the promoters assigned to these genes [9] by open boxes: directions of transcription from these promoters are noted by horizontal arrows. The plasmids pLB6 and pLB7 were derived from pLB3 by Bal 31 deletions starting from the Pvu II site: BamH I linkers ( ) were attached to the end points of the deletions. The plasmids conferred EcoR V restriction (r) and EcoR V modification (m) phenotypes as shown [9].

These studies indicated that the two genes for the EcoR V R/M system were transcribed divergently from the intergenic region: sequences with strong homology to the E. coli consensus promoter were found for each gene at the positions indicated in Figure 1. However, the DNA immediately preceding the translational starts had only weak homology to the Shine-Dalgarno sequence. Both open reading frames featured very high A + T contents (65%, ad compared to the E. coli [9] average of 49%) and contained a high proportion of codons that are infrequently used in E. coli [9]. Both the high A + T content and the rare codons were also observed in the EcoR I R/M genes [10], though there is no further similarity in either genetic organization or protein sequences between the EcoR I and EcoR V R/M systems. However, the genes for the Pst I system are organized as those for EcoR V [11].

The work of Bougueleret et al [9] also showed that the transcript from the gene for the EcoR V nuclease had the potential to form a very stable stem-loop structure at the 3' end of the coding sequence. A striking feature of the stem-loop is that the loop has an 8 base segment with perfect complimentary to a segment at the 5' end of the coding sequence. Thus the

3' and 5' segments could interact by base-pairing: such an interaction would impede the initiation of translation on the mRNA for the EcoR V nuclease. The role of the 3' segment was tested by deletion mutagenesis: its removal doubled the synthesis of the EcoR V nuclease [12]. Hence, a long-range interaction between two segments of mRNA separated by nearly 800 bases may regulate the production of EcoR V restriction enzyme, though other effects such as mRNA stability have yet to be excluded [12].

III. OVER-PRODUCTION AND PROTEIN PURIFICATION

Natural isolates of bacteria that produce type II restriction enzymes usually synthesize very little of these proteins. For example, the EcoR I restriction enzyme constitutes about 0.01% of the total protein in the natural strain of E. coli [4], and many other restriction enzymes are present in bacteria at even lower levels. Consequently, studies on the mechanism of action of restriction enzymes can be hampered by the difficulty in purifying the protein and by the low amount of pure protein that is finally obtained. Both problems can be overcome by cloning the gene for the enzyme in a vector that results in its enhanced expression. Strains that over-produce either the EcoR V restriction endonuclease or the EcoR V modification methylase have been constructed [13].

The constructions of both over-producers employed plasmid vectors that carried the P_L promoter from the phage λ and, downstream of the promoter, the transcription terminator from phage fd: the gene to be over-expressed was cloned between these sites. The recombinant DNA formed with these vectors was used to transform strains of E. coli that contained the temperature sensitive cI857 repressor from phage λ, encoded for by either a cryptic λ prophage in the chromosome or a separate plasmid (pRK248) [14]. This allowed the overproduction to be controlled: transcription from P_L promoter is blocked by the cI repressor at 28°C but the inactivation of the repressor at 42°C results in de-repression. Strains that were transformed by recombinants carrying the gene for the EcoR V nuclease also contained another plasmid (pLBM) which produced the EcoR V methylase from its endogenous promoter: consequently, incomplete repression of the gene for the nuclease at 28°C would not be lethal to the host [13].

For the EcoR V modification methylase, the BamH I - Hind III fragment from pLB7 (Figure 1) was cloned in the λP_L vector and the resultant transformants generated substantial amounts of the protein [13]. Within 3 h of temperature induction, more than 10% of the cellular protein in this strain was EcoR V methylase. For the nuclease, a similar construct was made by cloning the Mlu I - BamH I fragment from pLB6 (Figure 1) downstream of the λP_L promoter but strains carrying this plasmid yielded only a modest amount of EcoR V nuclease. Consequently, the configuration between gene and promoter had to be optimized. This was done by using nuclease BamH I to delete the majority of the DNA between the P_L promoter and the 5' end of the coding sequence [13]. The resulting recombinant, pTZ115, directed the synthesis of EcoR V nuclease to 2-5% of the cellular protein. A further improvement, pTZ115-14, was made by deleting sequences beyond the 3' end of the reading frame and this yielded 5-10% of the total protein as EcoR V restriction enzyme [12].

The over-production of the EcoR V restriction enzyme achieved to date cannot match the levels obtained with the EcoR I restriction enzyme [14]. For the latter, a strain constructed along very similar lines to that employed for EcoR V produced nearly 50% of its protein as EcoR I restriction enzyme. But in this over-producer, about 75% of the EcoR I protein formed an insoluble aggregate and no method was found for recovering active EcoR I from this aggregate [4]. In marked contrast, the purification of the EcoR V restriction enzyme from its over-producer recovered virtually all of the protein as active enzyme.

D'Arcy et al have described one method for purifying the EcoR V restriction endonuclease from the over-producing strain of Bougueleret et al [13]: chromatography on PC followed by gel filtration. As with many other restriction enzymes, PC was particularly effective at removing not only the nucleic acids but also the majority of other proteins. However, in our hands, gel filtration failed to separate the EcoR V enzyme from all of the impurities left after the PC column, despite some of the contaminants having much lower M_r values than EcoR V. The EcoR V restriction enzyme shows anomalous behavior in gel filtration (section IV). Rather than extending the purification to three columns, we modified the procedure of D'Arcy et al by replacing gel filtration with chromatography on Blue-agarose. The efficiency of Blue-agarose in the purification of restriction enzymes has been noted previously [15].

For the preparation described here, E. coli 1100 [pTZ115-14, pLBM, pRK248] [13] that had been previously tested for the presence of all three plasmids by selection with ampicillin, chloramphenicol and tetracycline, was grown at 28°C in 10 L of L-broth containing 100 µg/ml ampicillin. When the culture reached an A_{550} of 0.4, L-broth at 55°C (10 L) was added and subsequent growth was continued overnight at 42°C. The cells were harvested by centrifugation, washed and stored at -20°C: no EcoR V activity was lost during 3 months storage. This culture yielded 37 g cell paste containing 5.10^8 U EcoR V restriction enzyme. One U of EcoR V restriction activity was defined by the minimum amount of enzyme required to complete the cleavage of 0.5 µg λ DNA in 20 µl buffer R within 30 minutes at 37°C: buffer R is 50 mM

Tris, 100 mM NaCl, 10 mM $MgCl_2$, 10 mM βME, 0.01% (w/v) BSA, pH 7.5. The products of the reactions were analyzed by electrophoresis through agarose.

The cell paste was resuspended in 200 ml buffer X containing 0.8 M NaCl: buffer X is 20 mM K_2HPO_4, 10 mM βME, 1 mM EDTA, pH 7.0 but, for the initial extraction and the subsequent PC column, this was supplemented with the protease inhibitors, phenylmethylsulphonyl fluoride at 50 μM and benzamidine at 100 μM. This and all subsequent stages in the purification were carried out at 0-4°C. The cells were disrupted by sonication (6 x 2 min) and cell debris was removed by centrifugation (100,000 g for 0.5 hours). The supernatant from the centrifugation contained all of the EcoR V restriction enzyme initially present in the cells. However, if the initial extraction was carried out in the presence of <0.8 M NaCl, some of the EcoR V enzyme pelleted with the cell debris. Both of the columns in the subsequent purification required the NaCl concentration of the solution to be reduced from that in the previous stage of the preparation. This was done by diluting the enzyme with buffer X rather than by dialysis, because we have experienced major losses of the EcoR V enzyme during dialysis, presumably due to its sticking to dialysis membrane.

The extract (200 ml) was diluted with three volumes (600 ml) of buffer X, in order to reduce the NaCl to 0.2 M, and this was applied to a 40 ml column of PC (Figure 2). At this ionic strength, all of the nucleic acids and also the majority of the proteins did not stick to PC while virtually 100% of the EcoR V was retained. The restriction enzyme was eluted with a gradient of NaCl. Fractions from the gradient were analyzed by electrophoresis of the proteins through AP in the presence of SDS and by assays for EcoR V activity (Figure 2). The major protein of M_r 29,000, eluting predominantly between 0.5 and 0.6 M NaCl, coincided with the EcoR V activity and the peak fractions (70 ml) were pooled. At this stage, the EcoR V restriction enzyme was about 80% pure, the major contaminant being a small protein of M_r 12,000.

We tested a number of chromatography matrices for the removal of the contaminants left after the PC column, and found none superior to Blue-agarose (Figure 3). After diluting the peak fractions from the PC column with 20 volumes of buffer X, so as to reduce the NaCl concentration to 0.025 M, nearly all of the EcoR V stuck to a 20 ml column of Blue-agarose. Elution with a gradient of NaCl produced rather a broad peak of restriction enzyme (Figure 3). But, with the exception of minor contaminants at the leading edge of the peak (and consequently these fractions were discarded), no other proteins eluted in the same range of NaCl concentrations as the EcoR V enzyme. All contaminants had now been reduced to less than 0.5% of the total protein in the preparation. The pure fractions from the Blue-agarose column were pooled, added to an equal volume of cold glycerol and stored at -20°C. The preparation described here yielded 55 mg of EcoR V restriction endonuclease, with an activity of 4.10^8 U. Hence the specific activity of EcoR V is 7.10^6 U/mg, which is midway between the values seen with the EcoR I and SalG I restriction enzymes [4,17].

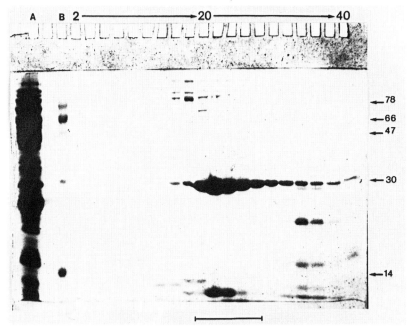

Figure 2. Purification of EcoR V on PC. The extract from the EcoR V overproducer (800 ml) was pumped onto a 1.6 x 20 cm column of PC (Whatman P11) that had been equilibrated in buffer X with 0.2 M NaCl. The flow rate was 40 ml/hour. The column was then washed with buffer X containing 0.2 M NaCl until no further protein eluted from the column, as judged by the A_{280} of the column eluate. Less than 1% of the EcoR V restriction activity eluted during either loading or washing stages. A linear gradient of 200 ml 0.2 M NaCl to 200 ml 0.8 M NaCl, both in buffer X, was then applied and the eluate at this stage collected in 40 fractions (10 ml each). Samples (20 μl) were analyzed by electrophoresis through PA (5% stack, 12% separating gel [16]) in the presence of SDS and the gel was stained with Coomassie Blue: lane A, the initial extract, lane B, protein standards, M_r values shown on the right of the gel; lanes 2→ 20→ 40, representing every other fraction from the gradient. Samples were also assayed for EcoR V restriction enzyme and fractions containing >2.106 U/ml are marked by the horizontal bar below the gel.

IV. PROTEIN STRUCTURE AND ENZYME MECHANISM

In the presence of SDS, the electrophoretic mobility of the EcoR V restriction enzyme through PA corresponds to a M_r value of 29,000 for the protein subunit (Figure 3). This agrees closely with that of 28,618 predicted from the DNA sequence [9]. However, two different values have been reported for the M_r of the native enzyme in solution. Kuzmin et al [18] found that filtration through Sephacryl S-300 yielded a M_r value of 26,000: this would be consistent with the EcoR V restriction enzyme being a monomer in solution. In contrast, D'Arcy et al [7] obtained M_r values of 70,000 from equilibrium sedimentation and 64,000 from the rotational correlation time of the protein as measured by fluorescence polarization. The latter method also gave a value of 71,000 for the EcoR V enzyme in the presence of an oligonucleotide, CCGATATCGG, that contains the EcoR V recognition site. The data indicated that EcoR V is a dimer in solution, both when free and when bound to its recognition site.

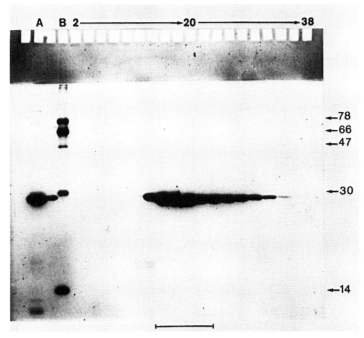

Figure 3. Purification of EcoR V on Blue-Agarose. The peak fractions from the PC column (70ml) were added to 1.4 L buffer X and this solution was pumped at 60 ml/hour onto a 1.6 x 10 cm column of Blue-agarose (Pharmacia Blue Sepaharos CL-6B) that had been equilibrated in buffer X with 0.025 M NaCl. After loading, the column was washed with 60 ml buffer X with 0.025 M NaCl: about 1% of the EcoR V activity was found in the eluate from the loading and wash stages. A linear gradient of 0.025 M NaCl to 0.4 M NaCl, bith in buffer X, was then applied at 20 ml/hour and the eluate was collected in 40 fractions (5 ml each). Samples (15 μl) were analyzed by electrophoresis through PA in the presence of SDS as in Figure 2: lane A, peak fractions from PC; lane B, protein standards, M_r values shown by the right of the gel; lanes 2→ 28, represent every other fraction from the gradient except n° 40. Samples were also assayed for EcoR V restriction enzyme and fractions containing >5.10^6 U/ml are marked by the horixontal bar below the gel.

To resolve this discrepancy, we have analyzed the apparent M_r of the EcoR V restriction enzyme by both gel filtration (Table I) and sedimentation velocity (Figure 4). Our experiments differ from those of D'Arcy et al [7] in that we used buffers in which the EcoR V enzyme is active, and also much lower concentrations of the protein (2 μg/ml instead of >100 μg/ml), so that our conditions approximate to assay conditions for EcoR V. In the presence of 100 mM NaCl, the optimum for EcoR V activity, gel-filtration yielded an apparent M_r of 30,000 for the EcoR V enzyme (Table I). This value agrees with that of Kuzmin et al [18] but other data in Table I shows that this cannot be taken as evidence for a monomeric structure for EcoR V. The apparent M_r values varied with the concentration of NaCl. For example, at 250 mM NaCl, the M_r value fell midway between those expected for the monomer and dimer of EcoR V: this apparent M_r was found to be independent of the protein concentration, thus excluding the possibility that it was due to rapid equilibration between monomeric and dimeric forms [19]. In contrast, at 50 mM NaCl, gel filtration yielded an apparent M_r substantially lower than that for the protein subunit.

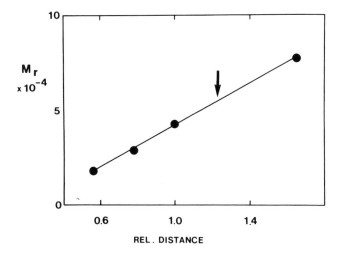

Figure 4. Sedimentation velocity of EcoR V. A sample of EcoR V endonuclease (0.02 μg in 0.1 ml also containing 100 μg ovalbumin) was centrifuged at 4°C through 4.6 ml 5-15% sucrose in 50 mM Tris, 100 mM NaCl, 6 mM $MgCl_2$, 10 mM βME, 0.01% BSA, pH 8.0 for 23 hours at 48,000 rpm. Marker proteins (myoglobin, carbonic anhydrase, ovalbumin, and ovotransferrin) were centrifuged in parallel through an identical gradient. Fractions were collected from the bottom of each gradient and analyzed for EcoR V activity or for marker proteins, the latter by electrophoresis through PA as in Figure 2. Data points (O) indicate the distances sedimented by each marker protein and the vertical arrow that for EcoR V: all distances are relative to that for ovalbumin.

TABLE I. Gel-filtration of EcoR V

NaCl (mM)	K_{av}	Apparent M_r
800	0.32	49,000
250	0.34	44,000
100	0.44	30,000
50	0.84	5,000
25	1.26	--

Samples of EcoR V restriction endonuclease (4.5 μg in 2 ml) were filtered through a 1.6 x 70 cm column of Sephacryl S-200 (Pharmacia) that had been equilibrated in 50 mM Tris, 6 mM $MgCl_2$, 10 mM βME, 0.01% (w/v) BSA, pH 8.9, with NaCl as indicated above. Fractions were collected and assayed for EcoR V activity. Elution volumes (V_e) were determined from the peak of activity and are expressed above as K_{av} values: K_{av} is defined as $(V_e - V_o)/(V_t - V_o)$ where V_t and C_o are the total and void volumes of the column, the latter being determined by the filtration of Blue-Dextran. Marker proteins of known M_r values were separately filtered through the same column and, for each, V_e was measured from the A_{280} of the filtrate. The relationship between K_{av} and M_r for the marker proteins (not shown) yielded the apparent M_r values for the EcoR V enzyme given above. (However, it is impossible to evaluate M_r from $K_{av}>1.0$)

One explanation for the data in Table I is that the flow of the EcoR V restriction enzyme is retarded by ionic interactions with the gel. The matrix of Sephacryl S-200 contains a number of acidic groups which can hinder the flow of a basic protein through the column [20]. The EcoR V enzyme is a basic protein: we found by isoelectric focussing that its pI was 9.4 (data not shown). At high ionic strength, electrostatic interactions between the protein and the gel would be weakened, which accounts for the increase in the apparent M_r of the EcoR V enzyme with increasing concentrations of NaCl (Table I). Though the majority of type II restriction enzymes characterized to date exist in solution as dimers or tetramers of identical protein subunits, a few have been reported on the basis of gel filtration experiments to be monomers [1]. However, as electrostatic forces play a role in many DNA-protein interactions, other proteins that bind to DNA might also interact with gel matrices and thus produce erroneous M_r values. We have observed that another type II restriction enzyme, Cau II, also behaves anomalously in gel filtration (S. P. Bennett and S. E. Halford, unpublished).

The sedimentation velocity of the EcoR V restriction enzyme through a sucrose gradient containing 100 mM NaCl is shown in Figure 4. Analysis of this data, by the method in [21], yielded an M_r of 55,000 for the EcoR V enzyme, in close agreement with that expected for the dimer. Hence this experiment not only supports the conclusions of D'Arcy et al [7] that the native form of the EcoR V enzyme is a dimer of identical protein subunits but also that the EcoR V enzyme remains a dimer at low concentrations of the protein. In the following discussion, molarities of the EcoR V enzyme are given in terms of the dimeric protein of M_r 57,000, the value being taken from the DNA sequence [9].

The EcoR V restriction endonuclease can cleave DNA under a wide range of experimental conditions. The standard buffer we employ for reactions of this enzyme on DNA is buffer R (section III). But the following variations to the pH or to the concentrations of either NaCl or $MgCl_2$ caused virtually no change to the activity of EcoR V on λ DNA, when the other components of buffer R were kept constant. The enzyme was equally active across the pH range from 7.0 to 8.5 and from 50 mM to 150 mM NaCl, though NaCl concentrations above 150 mM caused a major drop in activity. As with other type II restriction enzymes [1], the activity of EcoR V was absolutely dependent upon divalent cations, though altering the concentration of $MgCl_2$ from 1.0 mM to 20 mM had no effect upon the rate of the reaction and a number of other cations could substitute for Mg^{2+}. In the presence of 2 mM divalent metal chloride, the following reaction rates were obtained: Mg^{2+}, 100%; Co^{2+}, 40%; Ni^{2+}, 10%; Mn^{2+}, 5%. Buffer R also contained both βME and BSA because the EcoR V enzyme was inactivated by either modification of its cysteine (see below) or by dilution to very low protein concentrations.

To study the mechanism of action of the EcoR V restriction enzyme, we have used as the substrate the plasmid pAT153[22], a 3.6 kb derivative of pBR322 that contains one copy of the EcoR V recognition site. The monomeric form of this plasmid was isolated from a recA strain of E. coli and was obtained as a covalently closed circle of duplex DNA. If the EcoR V enzyme were to initially cleave only one strand of the DNA, the closed circle would be converted into the open circle form of the DNA: subsequent cutting of the intact strand of the open circle would finally yield the linear form of the DNA. Alternatively, if both strands were cut in a single concerted reaction, the closed circle of DNA would be converted directly into the linear form. The equivalent assay has been used previously on other restriction enzymes [1,2,23,24].

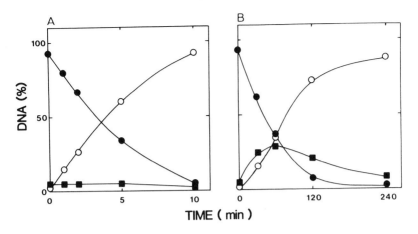

Figure 5. Reaction profiles for EcoR V.
A) The reaction contained 0.5 nM EcoR V restriction enzyme and 10 nM pAT153 (initially 94% covalently closed circle : the DNA had been labelled by the incorporation in vivo of [methyl-^{3}H] thymidine) in 100 μl buffer R at 37°C. Samples (20 μl) were taken from the reaction at the indicated times (zero time being before the addition of enzyme), quenched, and then subjected to electrophoresis through 1.2% agarose in order to separate the covalently closed (OO), open circle (□) and linear forms (O) of the DNA. The amount of DNA in each form was determined by scintillation counting and is recorded above as a % of the total DNA in each sample.
B) As A except the concentration of $MgCl_2$ in the buffer was 0.1 mM instead of 10 mM.

Figure 5a shows the reaction profile for the EcoR V restriction enzyme in its standard conditions, buffer R. The covalently closed DNA was converted directly to linear form. Hence, in this reaction, the enzyme has cleaved both strands of the DNA in a concerted reaction. Since the EcoR V enzyme is a dimer, this could be achieved by the formation of a symmetrical DNA-protein complex of the type observed with EcoR I [3]. Hence, one subunit of the EcoR V enzyme could be in position to cleave one strand of the DNA and the second subunit the other strand. However, when the reactions of EcoR V were studied at either low $MgCl_2$ concentrations (Figure 5b) or at pH values below 6.0 (not shown), open circle DNA was observed during the course of the reactions. The maximal concentration of the open circle form in the reaction shown in Figure 5b is much greater than the concentration of enzyme and thus the open circle DNA cannot be purely an enzyme bound intermediate of the reaction. Instead, under these sub-optimal conditions (the reaction in Figure 5b being much slower than that in Figure 5a), a significant portion of the DNA must be cleaved initially in only one strand of the duplex. Other reaction conditions that caused sub-optimal activity of the EcoR V enzyme, either too low or too high concentrations of

NaCl or high pH, failed to cause the production of the open circle form of pAT153 (not shown).

The reaction profiles of the EcoR V enzyme (Figure 5) differ from those obtained from the reactions of EcoR I on the plasmid pMB9 [2,23] but are very similar to those for SalG I [24]. It is striking that the experimental conditions that caused EcoR V to cleave just one strand of the DNA, either low $MgCl_2$ concentrations or low pH, also caused SalG I to do likewise. To account for this, we suggest that, in order to cut both strands of the DNA, both subunits of the EcoR V enzyme (or the SalG I enzyme) must bind Mg^{2+}: at low concentrations of $MgCl_2$, a fraction of the protein will have bound only one Mg^{2+} ion per dimer, and this fraction would cleave only one strand of the DNA. Moreover, it is possible that Mg^{2+} binds to acidic residues in the protein: the protonation of these residues at low pH would reduce their affinity for Mg^{2+}.

Very little is known at present about which side chains of the EcoR V protein interact with the substrate. Some chemical modification experiments have been carried out [19,25]. The EcoR V enzyme has one cystine/protein subunit [9]: modification of this group by N(1-pyrene) malemimide completely inactivated the enzyme. This contrasts with the EcoR I restriction enzyme, which also contains one cysteine/subunit but which can be labelled with N(1-pyrene) maleimide with full retention of activity [25]. The inactivation if EcoR V was not due solely to the large size of the pyrene substituent as N-ethylmaleimide and other thiol reagents also inactivated this enzyme [19]. Fluorescamine, a reagent that reacts preferentially with amino groups at alkaline pH values [27], was also tested on EcoR V (Figure 6). Not surprisingly, given the role of basic residues in DNA-protein interactions, even low levels of fluorescamine readily inactivated the enzyme. But when fluorescamine was used in the presence of the oligonucleotide, GGGATATCC, several residues could be modified without inactivating the protein (Figure 6). Presumably, the oligonucleotide had protected certain lysines that are required for enzyme activity. At acidic pH values, fluorescamine has been reported to modify proteins solely at the N-terminal amino group [27]. The reaction between fluorescamine and the EcoR V enzyme at pH 6 lead to the incorporation of only one fluorescamine/protein subunit and this conjugate had no activity [19]. This raises the possibility (which has yet to be proved) that the N-terminal amino group of the EcoR V enzyme, like that of the cI repressor [26], interacts with DNA.

However, it is probable that substantially more information on the structure of this protein will be available in the near future. The EcoR V restriction enzymes crystallizes readily: addition of polyethylene glycol

4000 to the EcoR V enzyme produced small crystals within 1 hour, which then grew to fill size within a few days [7]. Several different forms of the crystals were obtained of which two appear to be suitable for high resolution structure analysis. These grew to large sizes, diffracted X-rays to 2A resolution and contained one dimer of the protein per asymmetric unit [7]. The structure of the EcoR V enzyme has now been solved to 6A resolution: this revealed the overall shape of the protein, from which it was noted that only one face of the structure appeared to be a plausible site for contact with DNA [28]. Progress is being made on extending the resolution to 2.5A (F. Winkler, unpublished). Crystals of the EcoR V restriction enzyme have also been obtained in the presence of a series of self-complementary oligonucleotides [28]. These were stable in the absence of Mg^{2+} and were found to contain one duplex of the oligonucleotide per protein dimer. The oligonucleotides that were tested included some with the EcoR V recognition sequence (GATATC) but others had instead the EcoR I site (GAATTC): the latter is effectively non-specific DNA for, though the EcoR V enzyme can cleave DNA at the sequence GAATTC (Section V below), it does so at a very low rate. Only some of the co-crystals are suitable for X-ray diffraction studies: others form thin plates but these have been analyzed by image reconstruction from electron diffraction patterns [28].

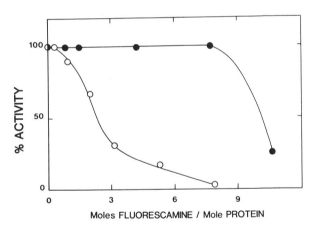

Figure 6. Fluorescamine modification if EcoR V. Reactions were between fluorescamine (0.01-1.2 mM) and EcoR V restriction enzyme (1.5 µM) at 20°C in 50 mM Pipes, 150 mM NaCl, 1 mM EDTA, 10% (v/v) glycerol, pH 7.5 in either the absence (O) or presence (●) of 20 µM GGGATATCCC. (The oligonucleotide would not have been cleaved by EcoR V, on account of the EDTA in the reaction.) The number of moles fluorescamine/mole EcoR V protein (dimer) was determined by comparing the flourescene of the dye-protein conjugate in 6M guanine-HCl, 50 mM borate, ph 8.5, with standards of fluorescamine-glycerine conjugate in the same buffer.

V. SPECIFICITY AND RELAXED SPECIFICITY

Type II restriction enzymes normally display very high specificities for their recognition sites on DNA: they cleave their recognition sites very much faster than any alternative DNA sequence [2]. We have examined the specificity of the EcoR V restriction enzyme on the same substrate, the plasmid pAT153, as used in the studies on its mechanism of action described above. This circular DNA molecule has a single recognition site for EcoR V so if the EcoR V enzyme were to cleave this DNA exclusively at the recognition site, the plasmid would be converted solely from circular to linear form. Any further cutting of the DNA would indicate that the EcoR V enzyme can cleave DNA at sequences distinct from its recognition site, and thus provide a measure of its specificity.

In the standard reaction conditions, buffer R at 37°C, 0.5 nM EcoR V restriction enzyme completed the linearization of 10 nM pAT153 within 10 minutes (Figure 5a). However, when much higher concentrations of the enzyme were used in 1 hour reactions under the same conditions, 200 nM EcoR V enzyme failed to cleave pAT153 at any sites apart from the recognition site and only at 600 nM enzyme were any additional cleavages observed (Figure 7). Even at the latter concentration, only a small fraction of the linearized pAT153 was cleaved at one or more additional sites (Figure 7). Hence, in buffer R, the EcoR V restriction enzyme must cleave its recognition site faster than any other individual sequence that is present on pAT153, by a factor of at least 20,000.

In marked contrast, the EcoR V restriction endonuclease cleaved pAT153 at numerous sites when the reaction was carried out in buffer S (Figure 7), a buffer of higher pH than buffer R and which also contained DMSO. Hence, the EcoR V enzyme, like many other restriction enzymes, alters its specificity in an altered reaction environment [18,29]. Alterations to the specificity of restriction enzymes were first observed with EcoR I, the relaxed specificity (EcoR I*) being observed in reactions at reduced ionic strength and elevated pH[30]. For the EcoR V restriction enzyme, we had arrived at the composition of buffer S by looking for EcoR V* activity over a range of reaction conditions [29]. As with the EcoR I* activity, EcoR V* was enhanced at pH values above 8.5, at low ionic strength and in the presence of organic solvents (DMSO yielding more EcoR V* activity than glycerol). The EcoR V* activity was produced by the same gene as the EcoR V restriction endonuclease [29].

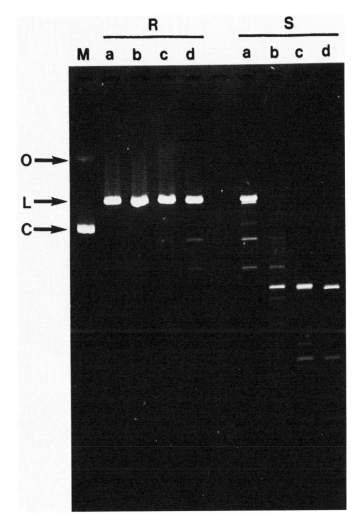

Figure 7. Specificity of EcoR V. Samples contained 10 nM pAT153 in either buffer R or buffer S, as indicated above the gel, and EcoR V restriction endonuclease at the following concentrations: lanes a, 20 nM; b, 60 nM; lanes c, 200 nM; lanes d, 600 nM. Buffer R was as in Section III: buffer S is 100 mM Tris, 10 mM $MgCl_2$, 10 mM β-ME, 0.01% (w/v) BSA, 10% (v/v) DMSO, pH 8.5. After 1 hour at 37°C, the DNA was analysed by electrophoresis through 1.2% agarose: the electrophoretic mobilities of the covanently closed (C), open circle (O) and linear (L) of pAT153 are marked on the left. Lane M: 0.5 μg pAT153.

With the EcoR I restriction enzyme, the altered specificity (EcoR I*) results in DNA cleavage at many sequences that differ by one or two bases from the canonical site for EcoR I, and a complicated hierarchy exists between susceptible and recalcitrant EcoR I* sites [31]. In contrast, the relaxed specificity of the EcoR V restriction enzyme yielded just two classes of EcoR V* sites, which we have called primary and secondary sites [29]. In buffer S at 37°C, cleavage of primary sites by 200 nM EcoR V enzyme was virtually complete within 1 hour (Figure 7) whereas extensive cutting of secondary sites was observed only in reactions with very high enzyme concentrations (2 μM) or when both the pH and the DMSO content of buffer S were increased (not shown here). Thirteen DNA fragments, between 1250 and 20 bp in size, were obtained by cleaving pAT153 at all of the primary EcoR V* sites: only the larger fragments are visible on the agarose gel shown in Figure 7, the smaller were detected on PA gels [29]. The EcoR V* activity also cleaved other DNA molecules, regardless of whether or not they contained the canonical site for EcoR V, into defined fragments: 7 from the 1.6 kb plasmid pAO3, 8 from SV40, and about 20 from ϕX174 RF DNA.

The DNA sequences recognized by EcoR V* were identified by first mapping, on these fully sequenced DNA molecules, the positions of several of the primary EcoR V* sites relative to the sites for other restriction enzymes [29]. For example, on pAT153, the EcoR V* sites adjacent to the BamH I, Hinc II, Hind III, Pst I and SalG I sites were all mapped to within ± 20 bp. The DNA sequences in the vicinity of each of the mapped sites were then inspected to see if they contained any features that might account for the specificity of EcoR V*. One of the EcoR V* sites on pAT153 was located at the position of the canonical site for EcoR V, GATATC. At all of the other primary sites, the DNA contained a sequence with 5/6 homology to the canonical site [29]. Table IIa provides a compilation of all of the sequences with 5/6 homology to the EcoR V site that were found at the locations of primary EcoR V* sites on either pAT153, pAO3 or ϕX174. However, the positions of the EcoR V* sites mapped on these DNA molecules excluded 6 out of the 18 possible sequences with 5/6 homology from being primary sites for the relaxed specificity of EcoR V (Table IIb) [29]. Direct evidence that the EcoR V* activity cleaves DNA at the sequences in Table IIa, but not (or only slowly) at the sequences in Table IIb, has been obtained by Alves and Pingoud (unpublished): sequencing of DNA fragments terminating at EcoR V* sties demonstrated, for example, that AATATC could be readily cleaved by EcoR V* whereas TATATC was resistant to cleavage.

Table II.

A) Sequences cleaved by EcoR V*	B) Sequences not cleaved by EcoR V*
G A T A T C	T A T A T C
	C " " " " "
A A T A T C	
	+G A G A T C
G G T A T C	
" C " " " "	+G A T C T C
" T " " " "	
	G A T A T A
G A C A T C	" " " " " G
" " A " " "	
G A T G T C	
" " " T " "	
G A T A C C	
" " " " A "	
" " " " G "	
G A T A T T	

A) Primary sites for the EcoR V* cleavage of pAT153, pA03 and φX174 RF DNA occurred on these DNA molecules at the locations of the sequences noted above.

B) All sequences with 5/6 homology to the canonical EcoR V site, but which are not primary sites for EcoR V*, are listed. All occur at least once on either pAT153, pA03 and φX174 RF DNA but the primary reaction of EcoR V* did not cleave the DNA at the locations of these sequences. However, two of these sequences, marked (+), were cleaved on DNA that lacked dam methylation.

There remains the possibility that other sequences, not found in Table IIa and which lack 5/6 homology to the canonical site, could also be primary EcoR V* sites. To test for this, the sizes of the fragments that would be obtained by cleaving DNA at all of the sequences in Table IIa were calculated for a series of fully sequenced DNA molecules. On pAT153, pA03 and φX174, the predictions from the DNA sequences agreed closely with the sizes of the fragments obtained in EcoR V* digests [29]. Hence no sequence present on any of these DNA molecules, other than those in Table IIa, constitutes a primary EcoR V* site.

Of the sequences in Table IIb that were not cleaved by EcoR V*, four differ from the canonical site for EcoR V by the replacement of either the guanine at the first position by a pyrimidine or the cytosine at the last position by a purine. Discrimination between purines and pyrimidines in DNA-protein interactions has been accounted for previously [31]. However, a different explanation accounts for why the remaining sequences in Table IIb, GAGATC and GATCTC, were not cleaved by EcoR V*: both contain the sequence GATC that is modified by the dam methylase of E. coli [32], and all of the DNA molecules used above were from dam^+ strains. This was confirmed by finding that the EcoR V* activity cleaved a sample of pAT153 from a dam^- strain of E. coli at more sites than pAT153 from the dam^+ strain, the extra sites being at the locations of these two sequences [29]. Likewise, SV40 DNA was cut by EcoR V* at the sequence GATCTC.

Therefore, the list of the primary EcoR V* sites in Table IIa must be supplemented by two sequences, GAGATC and GATCTC. However, DNA from E. coli is usually protected at these sites by dam methylation. It remains to be explained why buffer S should enhance the activity of the EcoR V restriction enzyme, relative to that in buffer R, against 14 out of the 18 sequences that differ from the canonical site by one base. Perhaps the DNA binding site of the EcoR V enzyme contains groups whose normal function in buffer R is to hinder the enzyme from interacting with any DNA sequence apart from the canonical site. Enhanced activity against non-canonical sites could then be due to altered DNA-protein contacts in buffer S, so that these blocking groups no longer fulfilled their function.

We have not analysed in full the secondary sites for the EcoR V* activity. However, we and Alves and Pingoud (unpublished) have found that two of the secondary sites appear to be located at the sequence TATATC and GAATTC. The former is one of the sequences with 5/6 homology that was excluded from being a primary site (Table IIb): while this is cleaved slowly by EcoR V*, no cutting at all could be detected at CATATC. The latter sequence, the EcoR I recognition site, differs from the canonical site for EcoR V by two nucleotides.

VI FUTURE PROSPECTS

The EcoR V restriction endonuclease is an attractive system for studies aimed at elucidating the molecular basis of DNA-protein interactions, for the following reasons.

i) The gene for this restriction enzyme has been characterized in depth, and its nucleotide sequence yields the amino acid sequence of the protein [9]. In addition, the availability of this gene on several different recombinant DNA molecules will facilitate further manipulation of this system, by techniques such as site-directed mutagenesis.

ii) Strains of E. coli have been constructed that over-produce the EcoR V restriction enzyme [12,13]. With these strains, large amounts of highly purified protein can be prepared readily (Figures 2 and 3).

iii) Crystals of the EcoR V restriction enzyme have been obtained as the free protein, the protein bound to its recognition sequence and the protein bound to non-specific DNA [7,28]. To date, this is the only protein that interacts with DNA for which crystals of all three forms have been described. The completion of X-ray diffraction analyses of both the protein and its complexes with DNA will thus reveal the structural basis for its recognition of, and discrimination between, DNA sequences.

iv) The EcoR V restriction enzyme readily cleaves self-complementary oligonucleotides that contain the EcoR V recognition sequence [7]. Hence, information about which groups on the DNA interact with the protein can be obtained by synthesizing oligonucleotides with modifications to either the bases, sugars or phosphate groups (B. Connolly; A. Pingoud; unpublished). Oligonucleotides containing bromodeoxyuridine have also been photochemically cross-linked to the EcoR V enzyme [33].

v) The EcoR V restriction enzyme, in marked contrast to EcoR I [4], is completely soluble in all solutions tested to date. Consequently, it will be possible to carry out biophysical studies on its interactions with DNA. For example, the fluorescence polarization of the EcoR V enzyme, labelled with fluorescamine under conditions that retain activity (Figure 6), undergoes a large change as it binds to DNA [25].

We thank Johan Botterman, Nigel Brown, Bernard Connolly, Alfred Pingoud, Fritz Winkler and Marc Zabeau for discussions and for unpublished data. We also thank Barbara Lovelady for her contributions to these experiments, and the Science and Engineering Research Council for financial support.

REFERENCES

1. Modrich, P. and Roberts, R.J. (1982) in Nucleases (Linn, S.M. and Roberts, R.J., eds.), Cold Spring Harbor, New York, pp 109-154.
2. Halford, S.E. (1983) Trends Biochem. Sci. 8, 455-460.
3. Frederick, C.A., Grable, J., Melia, M., Samudzi, C., Jen-Jacobson, L., Wang, B-C., Greene, P., Boyer, H.W. and Rosenberg, J.M. (1984) Nature 309, 327,331.
4. Luke, P.A. and Halford, S.W. (1985) Gene 37, 241-246.
5. Kholmina, G.V., Rebentish, B.A., Skoblov, Y.S., Mironov, A.A., Yankovskii, Y., Kozlov, Y.I., Glatman, L.I., Moroz, A.E., and Debabov, V.G. (1980) Dokl. Akad. Nauk SSSR 253, 495-497.
6. Schildkraut, L., Banner, C.D.B., Rhodes, C.S., and Parekh, S. (1984) Gene 27, 327-329.

7. D'Arcy, A., Brown, R.S., Zabeau, M., van Resandt, R.W., and Winkler, F.K., (1985) J. Biol. Chem. 260, 1987-1990.
8. Kessler, C., Neumaier, T.S., and Wolf, W. (1985) Gene 33, 1-102.
9. Bougueleret, L., Schwarzstein, M., Tsugita, A., and Zabeau, M. (1984) Nucl. Acids. Res. 12, 3659-3676.
10. Greene, P.J., Gupta, M., Boyer, H.W., Brown, W.E., and Rosenberg, J.M. (1981) J. Biol. Chem. 256, 2143-2153.
11. Walder, R.Y., Walder, J.A., and Donelson, J.E. (1984) J. Biol. Chem. 259, 8015-8026.
12. Botterman, J., De Almeida, E., Bougueleret, L., and Zabeau, M. (1985) in Sequence Specificity in Transcription and Translation, UCLA Symposium on Molecular and Cellular Biology: Vol. 21, Alan R. Liss, New York, pp. 397-407.
13. Bougueleret, L., Tenchini, M.L., Botterman, J. and Zabeau, M. (1985) Nucl. Acids. Res. 13, 3823-3839.
14. Botterman, J.H. and Zabeau, M. (1985) Gene 37, 229-239.
15. Baski, K., Rogerson, D.L., and Rushizky, G.W. (1978) Biochemistry 17, 4136-4139.
16. Laemmli, U.K. (1970) Nature 227, 680-685.
17. Maxwell, A. and Halford, S.E. (1982) Biochem. J. 203, 77-84.
18. Kuzmin, N.P., Loseva, S.P., Belyaeva, R.K., Kravets, A.N., Solonin, A.S., Tanyashin, V.I., and Baev, A.A. (1984) Mol. Biol. U.S.S.R. 18, 197-204.
19. Luke, P.A. (1986) Ph.D. Thesis, University of Bristol.
20. Belew, m., Porath, S., Fohlman, J., and Janson, J.C. (1978) J. Chromatogr. 147, 205-212.
21. Martin, R.G. and Ames, B.N. (1961) J. Biol. Chem. 236, 1372-1385.
22. Twigg, A.T. and Sherratt, D.J. (1980) Nature 283, 216-218.
23. Halford, S.E., Johnson, N.P. and Grinsted, J. (1979) Biochem. J. 179, 353-365.
24. Maxwell, A. and Halford, S.E. (1982) Biochem. J. 203, 85-92.
25. Luke, P.A. and Halford, S.E. (1986) Biochem. Soc. Trans. 14, in press.
26. Pabo, C.O., Krovatin, W., Jeffrey, A., and Sauer, R.T. (1982) Nature 298, 441-443.
27. Castell, J.V., Cervera, M. and Marco, R. (1979) Anal. Biochem. 99, 379-391.
28. Winkler, F.K., Brown, R.S., Leonard, K., and Berriman, J. (1986) in Crystallography in Molecular Biology, NATO Advanced Study Institute, Plenum, New York. In press.
29. Halford, S.E., Lovelady, B.M. and McCallum, S.A. (1986) Gene 41, 173-181.

30. Polisky, B., Greene, P., Garfin, D.E., McCurty, B.J., Goodman, H.M., and Boyer, H.W. (1975) Proc. Natl. Acad. Sci. U.S.A. 72, 3310-3314.
31. Rosenberg, J.M., Boyer, H.W., and Greene, P. (1981) in Gene Amplification and Analysis, Vol I: Restriction Endonucleases (Chirikjian, J.G. ed.), Elsevier, New York. pp 131-164.
32. Hattman, S., Brooks, J.E. and Masurekur, M. (1978) J. Mol. Biol. 126, 367-380.
33. Wolfes, H., Fliess, A., Winkler, F., and Pingoud, A. (1986) J. Biol. Chem. in press.

7

The Organization and Control of Expression of the *Pst* I
Restriction-Modification System

Roxanne Y. Walder and Joseph A. Walder

Department of Biochemistry
University of Iowa
Iowa City, IA 52242

I. INTRODUCTION

In Volume I of this series, we first reported the cloning and expression of the *Pst* I restriction-modification system in *Escherichia coli* [1]. The *Pst* I restriction enzyme and methylase genes were isolated from a genomic library of *Providencia stuartii* 164 DNA in *E. coli* HB101 on the basis of acquired resistance of the host to infection by bacteriophage lambda. Subcloning experiments localized the two genes to a 4.0 kilobase *Hin*d III fragment, the complete sequence of which has been determined [2,3]. In this report we review the organization and studies of the expression of the *Pst* I genes, and report the cloning of the *Pst* I restriction enzyme and methylase genes in yeast.

II. DETERMINATION OF THE NUCLEOTIDE SEQUENCE OF THE *PST*I SYSTEM

The DNA sequence of the 4.0 kb *Hin*d III fragment from *P. stuartii* 164 containing the restriction and modification enzyme genes revealed only two open reading frames of sufficient length to encode the two enzymes (see Figure 1) [3]. Analysis of deletion mutants and subcloning experiments showed that the longer open reading frame extending to the left corresponds to the methylase gene. The open reading frame on the opposite strand encodes the restriction enzyme. This arrangement precludes the possibility that the genes are transcribed as a polycistronic message. The two genes must be transcribed divergently from separate promoters. This same arrangement was later found in the *Eco*R V [4] and *Pvu* II systems [5]. In contrast, in the *Eco*R I [6,7], *Hha* II [8], *Bsu*R I [9], and *Pae*R7 [10] systems the restriction enzyme and methylase genes are colinear.

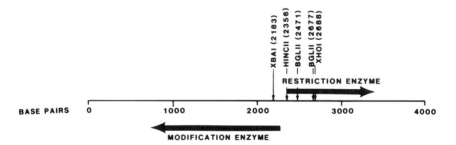

Figure 1. Map of the *Pst* I restriction and modification enzyme genes within the 4.0 kb *Hin*d III fragment isolated from *P. stuartii* 164. The open reading frames encoding the restriction enzyme and methylase genes are identified by the heavy black arrows. The positions of the *Xho* I, *Bgl* II, *Xba* I, and *Hin*c II restriction sites are shown.

III. THE PROTEINS OF THE *PST* I SYSTEM

The synthesis of the *Pst* I restriction and modification enzymes was studied in bacterial minicells and in a coupled in vitro transcription-translation system (Figs. 2 and 3).

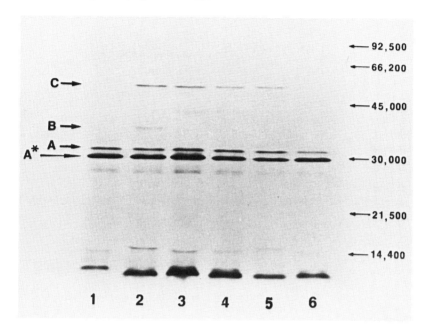

Figure 2. Polypeptides radiolabeled in *E. coli* minicells with L-[^{35}S] methionine. Bacterial minicells purified from X1411 transformed with different plasmids were labeled at 37°C for 90 minutes. Plasmids contained in minicells are as follows: Lane 1, pBR322; 2, pPst201; 3, pMe101 (*Xho* I - *Sal* I); 4, pMe201 (*Bgl* II - *Bgl* II); 5, pMe301 (*Bgl* II - *BamH* I); and 6, pMe401 (*Xba* I - *Sal* I). The restriction sites in parentheses indicate the boundaries of the DNA deleted from pPst201. Proteins were electrophoresed on a 10-18% SDS-polyacrylamide gel, followed by autoradiography. Arrows indicate the positions of (A) the β-lactamase precursor, (A*) the processed form of β-lactamase, (B) *Pst* I restriction endonuclease, and (C) *Pst* I methylase. The positions of molecular weight markers are also indicated.

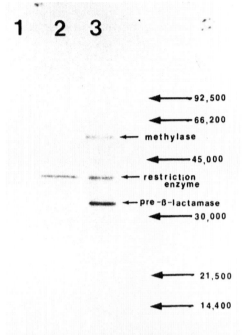

Figure 3. Immunoprecipitation of L-[^{35}S]methionine-labeled polypeptides synthesized in a coupled transcription- translation reaction programmed with pPst201. Lane 1, proteins reacted with preimmune IgG, 2, proteins precipitated with immune IgG against Pst I restriction endonuclease; 3, total proteins synthesized in the coupled system. The radiolabeled bands corresponding to the Pst I methylase, Pst I restriction enzyme and the precursor form of β-lactamase are labeled.

In minicells containing the plasmid pPst201, which includes the entire 4.0 kb Hind III fragment encoding the Pst I system, two new proteins were synthesized in addition to those encoded on the vector, pBR322 (see Figure 2). The smaller protein has an apparent molecular weight of 35,000 on SDS gels and co-migrates with the purified Pst I restriction enzyme. Deletions within the restriction enzyme gene in the plasmids pMe101, pMe201, and pMe301 result specifically in the loss of the synthesis of this protein (lanes 3-5). The larger protein has an apparent molecular weight of 58,000, compared to 56,830 calculated for the modification enzyme from the DNA sequence. In minicells containing the plasmid pMe401 (r-m-) the synthesis of this protein as well as the restriction enzyme was abolished (lane 6). The ratio of the methylase to the restriction enzyme in minicells transformed with the parent plasmid, pPst201, calculated from the densitometric trace of the autoradiogram after correction for the amount of methionine in the two proteins, is 4.8. As discussed below, the increased level of the methylase is probably related to an increased rate of synthesis rather than a decreased rate of degradation of the protein.

The amino acid sequence of the first 10 residues of the purified Pst I restriction endonuclease was determined by Edman degradation, and found to match precisely with that predicted from the DNA sequence beginning with the methionine codon within the open reading frame at nucleotide position 2396 (see Figure 4) [3]. This is the first methionine residue downstream of the transcription initiation site of the endonuclease gene (see below), and, therefore, must be the initiation codon. The only post-translational processing which occurs is the removal of the formyl group from the amino terminus. The deduced amino acid sequence indicates that the protein contains 326 amino acid residues and has a molecular weight of 37,370. The native molecular weight of the protein determined by gel filtration is 69,500 [11], indicating that the catalytically active form of the enzyme is a dimer.

Figure 4. Intergenic region and the beginning sequence of the Pst I restriction and modification enzyme genes. The transcription initiation sites at positions 2363 and 2292 for the restriction endonuclease and methylase, respectively, are circled and the direction of transcription, 5' to 3', is indicated by arrows. The sequences which best correspond to the -10 region for both transcripts are indicated by wavy lines. The region protected by RNA polymerase from DNase I digestion is enclosed by the box (see Figure 5). This corresponds to the expected site of the methylase promoter.

Previous studies indicate that at least a fraction of the Pst I restriction endonuclease is transported into the periplasmic space both in P. stuartii 164 and in E. coli [2,12]. We have not detected, even at the earliest stages of purification, a second peak of activity which might arise from proteolytic processing of the enzyme. Moreover the amino terminal sequence of the protein lacks the core of hydrophobic residues, typically 8 to 14 amino acids in length, characteristic of signal peptides

[13,14]. These observations suggest that the Pst I endonuclease is not further processed during membrane transport. The signal peptide may be located internally, as has been proposed for ovalbumin and several other secretory and integral membrane proteins [15-19].

In order to establish the translational start site of the modification enzyme, the protein was radiolabeled in minicells with either L-[^{35}S]methionine or L-[^{3}H]leucine, isolated by preparative polyacrylamide gel electrophoresis, and analyzed by Edman degradation [3]. Methionine residues were identified at cycles 1 and 39, and leucine residues at cycles 8 and 12 in agreement with the predicted amino acid sequence beginning with the methionine residue at nucleotide position 2265 (see Figure 4). Since this is the first methionine residue within the open reading frame it must be the initiation codon. The deduced amino acid sequence indicates that the protein contains 507 amino acid residues, and has a molecular weight of 56,830. By analogy with other Type II modification enzymes the catalytically active form of the enzyme is presumed to be a monomer. The site of methylation within the recognition sequence, CTGCAG, is at the adenosine residue [11].

The synthesis of both the restriction and modification enzymes was also observed in a coupled in vitro transcription-translation system programmed with pPst201 (Figure 3). The molecular weights of both proteins were identical to those detected in the minicell experiments, again arguing that neither is synthesized as a higher molecular weight precursor. In this case, the restriction enzyme was identified by immunoprecipitation with antibody raised against the purified protein (lane 2). Interestingly, the antibody did not cross-react with the methylase. The lack of homology which this suggests is also apparent from the analysis of the sequence of the two proteins. The longest stretch of identical amino acid residues is only a tetrapeptide, Glu-Arg-Ala-Leu, at amino acid positions 145-148 in the restriction enzyme and 164-167 in the methylase. Even allowing for conservative amino acid substitutions and for insertions or deletions within the sequence, there does not appear to be any significant homology between the two proteins. Similar comparisons between the restriction enzyme and methylase of the EcoR I [6,7], Hha II [8], BsuR I [9], EcoR V [4], and the PaeR7 [10] restriction-modification systems have also revealed little sequence homology between the two proteins.

IV. THE TRANSCRIPTION INITIATION SITES

The sites for initiation of transcription of the *Pst* I restriction enzyme and methylase genes were mapped using mung bean nuclease [3]. The 288 base pair *Xba* I-*Bgl* II restriction fragment, and the 505 base pair *Xba* I-*Xho* I fragment were used as templates (see Figure 1). The initiation site for transcription of the restriction enzyme gene is the adenosine residue at position 2363, 33 bases upstream from the initiation codon. The transcription of the methylase gene begins 27 bases upstream of its initiation codon at the adenosine residue at nucleotide position 2292. The two transcription initiation sites are separated by only 70 base pairs.

The nucleotide sequence in the intergenic region is shown in Figure 4. Sequences in the -10 region upstream from the transcription initiation sites that are homologous to the Pribnow box are underlined. For both genes, 4 of 6 residues match the consensus sequence 5'-TATAAT [20-23]. There is no apparent homology to the -35 region at the appropriate spacing, between 15 and 21 nucleotides [23], upstream from the putative Pribnow box of the endonuclease gene. The sequence 5'-TTATTA which occurs 17 bases upstream from the -10 region of the methylase gene shares three of six residues with the -35 consensus sequence. Although this degree of homology is not statistically significant, the extent of homology is similar in a number of known *E. coli* promoters.

Although the *Pst* I gene products are not highly abundant proteins in *P. stuartii* 164, the genes are much more efficiently expressed than in *E. coli*, by as much as 300 to 500-fold based on the level of the restriction enzyme [2]. DNA hybridization studies indicate that *Proteus* and *Providencia* species are only distantly related to other members of the Enterobacteriaceae family [24]. Consequently, the recognition sequences for RNA polymerase in *E. coli* and *P. stuartii* may be slightly different. In particular, the analysis of the *Pst* I genes suggests that the Pribnow box may be located further upstream from the transcription initiation site in Providencia species [3].

The initiation codon for both the restriction enzyme and methylase is the first AUG from the 5' end of the message. The nontranslated leader sequences are 33 and 27 nucleotides in length, respectively. Analysis of these sequences reveals no apparent homology with the Shine-Dalgarno sequence [25,26]. This again correlates with the fact that the genes are not highly expressed, but is probably not related to the difference in expression of the genes between *E. coli* and *P. stuartii* 164 given the very high degree to which the 3' end of the 16S rRNA is conserved [27].

V. DNASE I FOOTPRINTING THE THE METHYLASE GENE PROMOTER

Binding of *E. coli* RNA polymerase to the restriction enzyme and methylase promoters was studied by DNase I footprinting [3]. The *Xba* I-*Hin*d III fragment (residues 2183-3889) was used as the template (see Figure 1). This fragment contains the entire restriction enzyme gene and extends 108 bp downstream from the transcription initiation site for the methylase gene. The digestion of the fragment after incubation with 4 and 12 µg/ml *E. coli* RNA polymerase revealed a protected region between nucleotides 2267-2340, which occurs in the intergenic region (Figure 5). Roughly the same sequence was protected on the opposite strand when footprinting experiments were done on the *Xba* I-*Hin*d III fragment 3' end labeled with [α-32P]dCTP and *E. coli* DNA polymerase I. This region corresponds to the expected promoter site for the modification enzyme, spanning position -50 to +26 relative to the transcription initiation site (see Figure 4). The transcription initiation site for the endonuclease gene at position 2363 is clearly outside of this region. Even with a 5-fold higher concentration of *E. coli* RNA polymerase, there was no detectable binding to the restriction enzyme promoter.

217

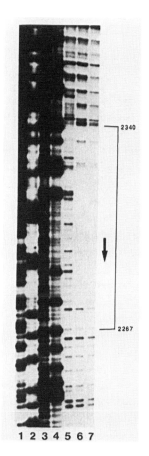

Figure 5. DNase I footprint of *E. coli* RNA polymerase binding to the 1.7 kilobase *Xba* I-*Hin*d III restriction fragment. The region protected from DNase I digestion, position 2267-2340, is bracketed. The fragment was 5'-end labeled at the *Xba* I site, followed by cleavage with *Hin*d III. From left to right, the first four lanes (overexposed) are the cleavage products of the Maxam-Gilbert sequencing reactions of the template. Reactions are A>C, G, T+C, and C, respectively. The next three lanes are from the footprinting experiment, where radiolabeled DNA template (90 ng) was incubated with different concentrations of *E. coli* RNA polymerase: lane 5) no polymerase, 6) 200 ng, and 7) 600 ng, followed by digestion with DNase I.

VI. CONTROL OF EXPRESSION OF THE *PST* I GENES

For restriction-modification systems which can be transformed into an unmodified recipient strain at high efficiency, coordination of the expression of the restriction and modification enzymes would appear to be essential. Unless the modification enzyme is expressed first to protect the DNA of the host, synthesis of the restriction enzyme would be lethal. For the phage P1-encoded restriction-modification system it has been shown that modification is expressed in advance of restriction in newly infected cells [28]. It is unclear why this mechanism would have evolved for restriction-modification systems that are chromosomally located, such as *Hha* II and *Pst* I. One possibility is that these two systems may have arisen from a recent chromosomal integration of a mobile genetic element carrying the restriction enzyme and methylase genes.

The initial model proposed for sequential expression of restriction-modification systems required the methylase to act as a positive regulator for expression of the endonuclease gene [29, 30]. Further studies of the *Eco*R I, *Hha* II and *Pst* I systems have ruled out this possibility [31, 32]. In the case of the *Pst* I system, we have established directly that the restriction enzyme can be expressed in the absence of the methylase in a cell-free transcription-translation system using a fragment containing only the restriction enzyme gene as the template [11].

The alternative gene arrangements in restriction-modification systems suggest different mechanisms of regulation. In the *Eco*R I and *Hha* II systems, the two genes are arranged in tandem on the same strand, and are very closely linked [6,7]. In the *Hha* II system, the methylase gene is 5' to the endonuclease. Cotranscription of the two genes in the order of the methylase to the restriction enzyme is apparently sufficient to ensure the sequential expression of the two proteins so that the cellular DNA is modified and protected before being cleaved by the endonuclease. In the *Eco*R I system the order of the genes is reversed: the endonuclease precedes the methylase. When the lac promoter is placed upstream of the endonuclease gene, the two genes are transcribed as a polycistronic message [31], and presumably are co-transcribed from the endogenous promoter of the endonuclease. There is, in addition, a separate internal promoter for the methylase gene allowing for its expression in the absence of the restriction enzyme [31]. Early synthesis from this promoter may be important in modifying the host DNA.

In the *Pst* I system, the restriction enzyme and methylase genes are encoded on opposite strands and hence must be transcribed divergently from separate promoters. DNase I protection experiments indicate that RNA polymerase forms a more stable complex with the methylase promoter (Figure 5). There was no detectable binding to the promoter of the restriction enzyme gene even with increasing concentrations of the polymerase, suggesting that the occupancy of each of the two sites is mutually exclusive. Since the transcription initiation sites for the restriction enzyme and methylase genes are separated by only 70 base pairs (see Figure 4), the two promoters probably overlap directly - the contact sites with *E. coli* RNA polymerase generally extend 40 to 50 bases upstream of the initiation site [33]. Preferential binding of RNA polymerase to the methylase promoter, would thus inhibit transcription of the endonuclease gene. In agreement with this conclusion, the minicell experiments shown in Figure 2 suggest that the methylase is more highly expressed than the restriction enzyme. The differential rate of synthesis of the two proteins is apparently sufficient to ensure that the DNA of the host is modified in newly transformed cells. As in other restriction-modification systems studied, there does not appear to be an obligate mechanism requiring that all cleavage sites within the chromosome are methylated before the restriction enzyme is expressed. The observation of site-specific recombination mediated by the *Eco*R I endonuclease suggests that to some extent restriction and religation of the chromosomal DNA does take place initially after transformation as well as during normal growth [34]. This potential recombinogenic activity may be an important function of restriction systems in addition to the elimination of foreign DNA.

VII. CLONING THE *PST* I GENES IN YEAST

Although nucleases which have some sequence specificity have been described in several mammalian species [35], there is not as yet evidence of a true restriction-modification system in higher eukaryotes. In *Chlamydomonas*, a lower eukaryote, the coordination of restriction and modification activities appears to be responsible for the maternal inheritance of chloroplast DNA [36]. In yeast, little if any DNA methylation occurs [37]. Against this background, the expression of restriction and modification activities might serve as a useful probe of the accessibility of different chromosomal locations, and of the activity of different recombination and repair systems.

Several bacterial genes have been cloned and expressed in yeast [38-43]. In our initial efforts to clone the *Pst* I genes in yeast, the 4.0 kb *Hin*d III fragment containing the entire *Pst* I system was inserted into the yeast shuttle vector YEp13 [44]. This plasmid contains sequences from the yeast 2u plasmid necessary for replication in yeast, the yeast leu 2 gene, and pBR322. The resulting plasmids, pYE002 and pYE004, having the insert cloned in opposite orientations, were shown to express both the restriction enzyme and methylase in *E. coli* HB101, and in the in vitro transcription-translation system described in Figure 3. Both plasmids transformed *Saccharomyces cerevisiae* strain LL20, a leu 2 auxotroph [45], at high frequency and were stably maintained (see Table I). The presence of the correct plasmid within the yeast cells was verified by Southern hybridization. Yeast cells carrying these plasmids, however, did not express detectable levels of either the restriction enzyme or methylase. Restriction digestion of DNA isolated from the transformed cells indicated that the *Pst* I sites within the plasmid as well as the chromosomal DNA were not modified. Permeabilized cell extracts did not contain either *Pst* I restriction enzyme or methylase activities, nor did large scale yeast cell extracts obtained by glass bead disruption. Evidently the endogenous promoters of the *Pst* I genes are not efficiently utilized in *S. cerevisiae*.

TABLE I. Transformation efficiencies of LL20 with plasmid DNA.

Plasmid	Transformants/μg plasmid DNA
YEp13	3000
pYE002	3300
pYE004	3100
AAH5	1900
pYE205	1200
pYE320	3
pYE340	1500

In order to obtain higher levels of expression of the *Pst* I genes, they were linked to a strong yeast promoter. For this purpose we utilized the plasmid AAH5, a derivative of YEp13 containing the promoter and termination elements of the yeast ADCI gene (kindly supplied to us by Benjamin Hall) [46]. *Hinc*II fragments of the 4.0 kb *Hin*d III insert of pPst201 containing either the restriction or modification enzyme gene were inserted into the *Hin*d III site within the 5' non-translated leader sequence of the ADCI gene. The resultant plasmids are diagrammed in Figure 6. The *Hinc* II site within the *Pst* I locus is located only 7 nucleotides upstream from the transcription initiation site of the endonuclease gene.

Cleavage at this site destroys the endonuclease promoter and prevents expression of the gene in *E. coli*. This allows the restriction enzyme gene to be maintained in *E. coli* in the absence of the methylase. The plasmid pYE340 has the restriction enzyme gene inserted in the opposite orientation from pYE320, such that transcription from the ADCI promoter would not result in expression of the restriction enzyme.

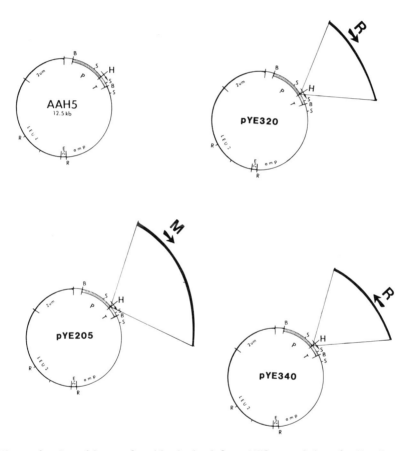

Figure 6. Recombinant plasmids derived from AAH5 containing the *Pst* I restriction enzyme or methylase genes. *Hin*c II-*Hin*d III fragments containing either the *Pst* I restriction or modification enzyme genes (see Figure 1) were inserted at the *Hin*d III site of AAH5, 12 nucleotides upstream of the initiator methionine of the ADHI gene. The directionality of the *Pst* I genes is indicated by heavy arrows and the restriction and modification enzyme genes are labeled as R and M, respectively. P=ADCI promoter, T=ADCI terminator, B=*Bam*H I, R=*Eco*R I, S=*Sph* I, H=*Hin*d III. The cross-hatched region indicates the ADCI promoter and terminator sequences.

The ADC1 promoter allowed expression of both the *Pst* I restriction enzyme and methylase genes in yeast. Expression of the methylase in cells transformed with the plasmid pYE205 was detected by Southern hybridization of yeast DNA isolated from transformed clones (Figure 7). Nick-translated pPst201 DNA was used as the probe. The original plasmid AAH5 contains two *Pst* I sites. Cleavage of this plasmid, when isolated from the yeast strain LL20, with *Pst* I gives rise to the two fragments of the expected sizes (lanes 1 and 2). These same two sites are present in pYE205. Analysis of DNA from yeast cells transformed with pYE205 showed that the plasmid DNA could not be cleaved by *Pst* I (lane 3), indicating that the modification enzyme was expressed and able to fully methylate the plasmid. In contrast, yeast chromosomal DNA isolated from transformants containing pYE205 appeared to be fully susceptible to cleavage by *Pst* I, when compared with DNA isolated from cells transformed with AAH5, at the level detectable by ethidium bromide staining of DNA fragments separated by agarose gel electrophoresis. This result indicates that the bulk of the *Pst* I sites within the chromosomal DNA of cells transformed with pYE205 were not modified. Similar results have been observed with the cloning of the *Bsp* I methylase and the *E. coli dam* methylase in *S. cerevisiae* [43, 47]. The difference in the levels of plasmid and chromosomal DNA methylation observed in these systems cannot be accounted for by the fact that the plasmid DNA when originally introduced into the yeast cell is already fully methylated. The *dam* methylase shows only a slight kinetic preference for hemimethylated substrates [48], and the methylases of Type II restriction-modification systems appear to be even less selective [49]. Recently, M.F. Hoekstra and R.E. Malone (personal communication) have shown that the chromosomal DNA of *rad*1 and *rad*3 excision repair mutants containing the *dam* gene become very highly or fully methylated. This result suggests that the low steady-state level of chromosomal DNA methylation observed in wild type strains is due to removal of the modified base (6-methyl-adenosine) by the excision repair system. Simple replicative plasmids derived from the 2μ circle are presumably less susceptible to this pathway. In this regard it will be of interest to determine the extent of methylation of plasmids which contain other chromosomal elements such as functional centromere and telomere sequences.

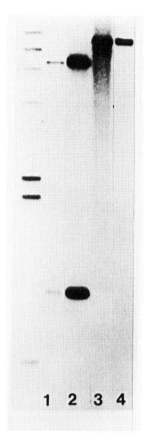

Figure 7. Southern blot analysis demonstrating expression of the *Pst* I methylase in yeast cells transformed with the plasmid pYE205. DNA was digested with *Pst* I, electrophoresed on a 0.7% agarose gel and transferred to nitrocellulose. The hybridization probe was nick translated pPst201. Lane 1, LL20 transformed with AAH5; 2, purified AAH5 plasmid from *E. coli* HB101; 3, LL20 transformed with pYE205; 4, purified pYE205 from HB101. Marker lanes on the left are end-labeled *Hin*d III fragments of lambda DNA 23, 9.4, 6.7, 4.4, 2.3, 2.0, and 0.6 kb in length.

The yeast ADCI promoter also appeared to allow expression of the *Pst* I restriction endonuclease in LL20. As shown in Table I, the transformation efficiency of the plasmid pYE320 is decreased by about five hundred-fold compared to either AAH5, the r-m+ plasmid pYE205, or pYE340, in which the endonuclease gene is inserted in the opposite orientation from that in pYE320 such that transcription of the gene can not occur from the yeast ADCI promoter. In two of the three clones that were obtained after transformation with pYE320, there were no DNA sequences related to the plasmid detectable on Southern blots probed with pPst201, and in the third clone substantial deletions of the plasmid had occurred. These results suggest that the endonuclease was expressed in cells transformed with the original plasmid, and in the absence of the methylase was lethal to the host.

The plasmid pYE320 may be useful as a positive-selection vehicle for cloning foreign genes into yeast. Because the endogenous promoter of the endonuclease gene has been deleted, the plasmid can be maintained in *E. coli* in the absence of the methylase gene. In yeast, however, expression of the *Pst* I restriction enzyme is possible by transcription off of the ADCI promoter and is lethal to the cell, as reflected by the low efficiency of transformation of LL20 with pYE320. Insertion of foreign DNA into the *Pst* I restriction enzyme gene or within the 5' regulatory region would prevent synthesis of the enzyme and allow the plasmid to be stably maintained in yeast.

VIII. REFERENCES

1. R.Y. Walder, J.L. Hartley, J.E. Donelson and J.A. Walder in: Gene Amplification and Analysis, Vol. 1, J.G. Chirikjian, ed. (Elsevier North-Holland, New York 1981) pp.217-227.
2. R.Y. Walder, J.L. Hartley, J.E. Donelson and J.A. Walder, Proc. Nat. Acad. Sci. 78, 1503-1507 (1981).
3. R.Y. Walder, J.A. Walder and J.E. Donelson, J. Biol. Chem. 259, 8015-8026 (1984).
4. L. Bougueleret, A. Tsugita and M. Zabeau, Nuc. Acids Res. 12, 3659-3767 (1984).
5. R.M. Blumenthal, S.A. Gregory and J.S. Cooperider, J. Bact. 164, 501-509 (1985).
6. A.K. Newman, R.A. Rubin, S.-H. Kim and P. Modrich, J. Biol. Chem. 256, 2131-2139 (1981).
7. P.J. Greene, M. Gupta, H.W. Boyer, W.E. Brown and J.M. Rosenberg, J. Biol. Chem. 256, 2143-2153 (1981).
8. B. Schoner, S. Kelley and H.O. Smith, Gene 24, 227-236 (1983).
9. A. Kiss, G. Posfai, C.C. Keller, P. Venetianer and R.J. Roberts, Nuc. Acids Res. 13, 6403-6421 (1985).
10. G. Theriault, R.H. Roy, K.A. Howard, J.S. Benner, J.E. Brooks, A.F. Waters and T.R. Gingeras, Nuc. Acids Res. 13, 8441-8461 (1985).
11. R.Y. Walder, Ph.D. thesis, The University of Iowa (1984).

12. D.I. Smith, F.R. Blattner and J. Davies, Nuc. Acids Res. 3, 343-353 (1976).
13. D.D. Sabatini, G. Kreibich, T. Morimoto and M. Adesnik, J. Cell Biol. 92, 1-22 (1982).
14. D. Perlman and H.O. Halvorson, J. Mol. Biol. 167, 391-409 (1983).
15. V.R. Lingappa, J.R. Lingappa and G. Blobel, Nature (Lond.) 281, 117-121 (1979).
16. W.A. Braell and H.F. Lodish, Cell 28, 23-31 (1982).
17. E. Santos, H. Kung, I.G. Young and H.R. Kaback, Biochem. 21, 2085-2091 (1982).
18. P.B. Wolfe, W. Wickner and J.M. Goodman, J. Biol. Chem. 258, 12073-12080 (1983).
19. R.R. Kopito and H.F. Lodish, Nature (Lond.) 316, 234-238 (1985).
20. D. Pribnow, Proc. Nat. Acad. Sci. 72, 784-788 (1975).
21. M. Rosenberg and D. Court, Ann. Rev. Genet. 13, 319-353 (1979).
22. U. Siebenlist, R.B. Sampson and W. Gilbert, Cell 20, 269-281 (1980).
23. D.K. Hawley and W.R. McClure, Nuc. Acids Res. 11, 2237-2255 (1983).
24. D.J. Brenner, J.J. Fermer III, G.R. Fanning, A.G. Steigerwalt, P. Klykken, H.G. Wathen, F.W. Hickman and W.H. Ewing, Int. J. Syst. Bact. 28, 269-282 (1978).
25. J. Shine and L. Dalgarno, Proc. Nat. Acad, Sci 71, 1342-1346 (1974).
26. G.D. Stormo, T.D. Schneider and L.M. Gold, Nuc. Acids Res. 10, 2971-2996 (1982).
27. G.W. Fox, E. Stackebrandt, R.B. Hespell, J. Gibson, J. Maniloff, T.A. Dyer, R.S. Wolfe, W.E. Balch, R.S. Tanner, L.J. Magrum, L.B. Zablen, R. Blackemore, R. Gupta, L. Bonen, B.J. Lewis, D.A. Stahl, K.R. Luehrsen, K.N. Chen and C.R. Woese, Science 201, 457-463 (1980).
28. W. Arber in: Progress in Nucleic Acids Research and Molecular Biology, Vol. 14, W.E. Cohn, ed. (Academic Press, New York 1974) pp. 1-34.
29. M.B. Mann, R.M. Rao and H.O. Smith, Gene 3, 97-112 (1978).
30. H.O. Smith, Science 205, 455-462 (1979).
31. C.D. O'Connor and G.O. Humphreys, Gene 20, 219-229 (1982).
32. M.B. Mann in: Gene Amplification and Analysis, Vol. 1, J.G. Chirikjian, ed. (Elsevier-North Holland, New York 1981) pp. 229-237.
33. A. Schmitz and D.J. Galas, Nuc. Acids Res. 6, 111-137 (1979).
34. S. Chang and S.M. Cohen, Proc. Nat. Acad. Sci. 74, 4811-4815 (1977).
35. W.G. McKenna, J.J. Maio and F.L. Brown, J. Biol. Chem. 256, 6435-6443 (1981).

36. R. Sager and R. Kitchin, Science 189, 426-433 (1975).
37. S. Hattman, J.E., Brooks and M. Masurekar, J. Mol. Biol. 126, 367-380 (1978).
38. C.P. Hollenberg in: Extrachromosomal DNA, ICN-UCLA Symposia on Cellular and Molecular Biology, Vol. 15, D. Cummings, P. Borst, S. Dawid, S. Weissman and C.F. Fox, eds. (Academic Press, New York 1979) pp. 325-338.
39. J.D. Cohen, T.R. Eccleshall, R.B. Needleman, H. Federoff, A.B. Buchferer and J. Marmur, Proc. Nat. Acad. Sci. 77, 1078-1082 (1980).
40. J.J. Panthier, P. Fournier, H. Heslot and A. Rambach, Curr. Genet. 2, 109-113 (1980).
41. A. Jiminez and J. Davies, Nature (Lond.) 287, 869-871 (1980).
42. R. Roggenkamp, B. Kustermann-Kuhn and C.P. Hollenberg, Proc. Nat. Acad. Sci, 78, 4466-4470 (1981).
43. Z. Feher, A. Kiss and P. Venetianer, Nature (Lond.) 302, 266-268 (1983).
44. J.R. Broach, J.N. Strathern and J.B. Hicks, Gene 8, 121-133 (1979).
45. J.W. Szostak and R. Wu, Plasmid 2, 536-554 (1979).
46. G. Ammerer in: Methods in Enzymology, Vol. 101, R. Wu, L. Grossman and K. Moldave, eds. (Academic Press, New York 1983) pp. 192-201.
47. J.E. Brooks, R.M. Blumenthal and T.R. Gingeras, Nuc. Acids Res. 11, 837-851 (1983).
48. G.E. Geier and P. Modrich, J. Biol. Chem, 254, 1408-1413 (1979).
49. P. Modrich, CRC Crit. Rev. Biochem. 13, 287-323 (1982).

8

The *Pvu* II Restriction-Modification System:
Cloning, Characterization and Use in Revealing an
E. coli Barrier to Certain Methylases or Methylated DNAs.

Robert M. Blumenthal

Department of Microbiology
Medical College of Ohio
Toledo, Ohio 43699

I. INTRODUCTION

The physiology of the type II restriction-modification systems (RMS2s) is not yet understood in detail. Relatively few RMS2s have actually been demonstrated to carry out the role for which they were named: restriction of DNA entry into the cell through selective endonuclease action [1]. On the other hand, there are numerous examples of DNA methylases, biochemically identical to RMS2 enzymes but apparently lacking a cognate endonuclease activity, for which no role has yet been discovered [2]. Our limited knowledge in this area has hindered attempts to understand why the genes for certain RMS2s have been difficult to clone and express (see reference 3). Several basic questions remain to be answered regarding the regulation of and sequence recognition by RMS2 enzymes. Also, does the methylase of a newly introduced RMS2 have to fully protect the DNA of its new host before the endonuclease gene is expressed, or can the cell repair a significant amount of endonucleolytic damage? And what are the effects of a foreign DNA methylase on the cell?

We have cloned and characterized the genes for the Pvu II RMS2 from Proteus vulgaris and, in the course of studies designed to examine their regulation, found that the majority of E. coli strains tested are poorly transformed by these genes. We found that this effect is due to the Pvu II methylase, not to the endonuclease, and that DNAs methylated by the Pvu II methylase also transform most tested strains poorly. Others have found this effect with several other DNA methylases, and it has raised new questions about gene cloning procedures and about E. coli physiology.

II. THE Pvu II RESTRICTION-MODIFICATION SYSTEM

Proteus vulgaris is a Gram-negative bacterium, in the family Enterobacteriaceae and tribe Proteae. Four type II restriction-modification systems (RMS2s) have been found among species of this tribe (Table I); two of them in Proteus vulgaris. [Tribe Proteae includes Proteus, Providencia, and Morganella.]

TABLE I. RMS2s of the tribe *Proteae*

Species	RMS2	Substrate	Reference
Proteus vulgaris	*Pmy* I	CTGCAG	*
Proteus vulgaris	*Pvu* I	CGAT↑CG	4
Proteus vulgaris	*Pal* I	CAG↑CTG	4
Providencia alcalifaciens	*Pst* I	GGCC	5
Providencia stuartii	*Pst* I	CTGCA↑G	6,7
Pvu I/*Pvu* II consensus	----	CRRYYG	
Pmy I-*Pst* I/*Pvu* II consensus	----	$C(^A_T)GC(^T_A)G$	

R, purine; Y, pyrimidine. *K. Weule and R.J. Roberts, unpubl. obs.

A number of interesting questions can be asked about these RMS2s. First, how are they regulated? Is there coordinate regulation of two RMS2s that naturally occur in the same cell (e.g., *Pvu* I and *Pvu* II)? What is the evolutionary relationship between these RMS2s? Are the related substrate sequences of *Pvu* I, *Pvu* II, and *Pst* I significant in this regard? How do restriction enzymes recognize their substrate sequences? To address the last question, this laboratory has been studying the family of RMS2s that include AGCT as an element of their substrate sequences (*Alu* I, *Hin*d III, *Pvu* II, *Ban* II, *Sac* I, *Hgi*A I, and *Nsp*B II). We have cloned and characterized the structural genes for the *Pvu* II system.

A. Cloning pPvu1

As a number of RMS2s are known to be plasmid coded, we have been screening the strains producing AGCT-family RMS2s for the presence of plasmids. Our rationale is that RMS2s coded for by mobile genetic elements might have evolved regulatory mechanisms consistent with such mobility, and be more easily cloned. A plasmid was found in one of the strains, *Proteus vulgaris* (ATCC 13315) [8].

We first detected the small, low copy-number plasmid in whole DNA preparations resolved on 1.0% agarose gels. We prepared a cleared lysate from a 4-liter culture of *P. vulgaris*, subjected it to isopyknic centrifugation in the presence of ethidium bromide, and harvested the region in which marker superhelical DNA (pBR322) had banded in a parallel gradient. From the 4 liters of culture, we obtained 8 µg of plasmid DNA. We refer to this plasmid as pPvu1, since it was the first plasmid we found in a strain of *P. vulgaris*.

This plasmid appears to carry the genes for the *Pvu* II, but not the *Pvu* I, RMS2 [8]. This was determined by ligating *Hin*d III or *Eco*R I-digested pPvuI into compatibly-digested pBR322, and transforming *E. coli* strain HB101. The pBR322 itself carries substrate sites for both *Pvu* I and *Pvu* II [9], and we found that all of our primary recombinant plasmids were resistant to *Pvu* II but were cleaved by *Pvu* I endonuclease. This implied that *Pvu* II methylase was being produced by cells harboring these recombinants, which we confirmed directly by showing that their extracts could protect bacteriophage λ DNA from subsequent digestion with *Pvu* II endonuclease [8]. For some of the recombinants, we could also demonstrate *Pvu* II endonuclease activity in extracts, and these same recombinant strains restricted the growth of λ*vir* bacteriophage 10^6 fold.

The *Pvu* I RMS2 genes are either not expressed in *E. coli* or are not carried by pPvuI. They may reside on the *P. vulgaris* chromosome. Type III restriction-modification systems are carried by some temperate bacteriophages [10], and many of the bacteriophages carry RMS2-like methylase genes [11]. In this light it is interesting that *P. vulgaris* appears to carry a prophage, which spontaneously forms plaques in *P. vulgaris* lawns at an extremely low frequency (unpublished observations). This is similar to what has been seen in *Proteus mirabilis* [12-14].

B. Relative Positions of the *Pvu* II Genes

The primary cloning yielded three different recombinant plasmids: pPvuM1.9, pPvuRM5.1, and pPvuRM5.2. They are named in accordance with the recommendations of Gingeras and Brooks [15]. Thus, for example, pPvuRM5.1 carries both the endonuclease (R) and methylase (M) genes of *Pvu* II on a roughly 5.1 kilobase insert. Restriction maps of these three primary recombinant plasmids are shown in Figure 1.

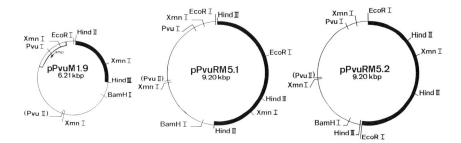

Figure 1. Restriction maps of the primary clones of pPvuI. The thick portions of the circles represent DNA from pPvuI; the remainder is derived from pBR322. The position of the *amp* gene is shown for pPvuM1.9, but all three plasmids carry this gene. The *Pvu* II site is methylated in all three plasmids, and is therefore enclosed in parentheses. kpb - kilobase pairs.

A subclone of pPvuRM5.1, named pPvuRM3.4, was isolated by removing the 1.7 kb segment between the pBR322- and pPvuI-derived EcoR I sites. As implied by its name, cells carrying pPvuRM3.4 were found to produce Pvu II methylase and endonuclease activities. This subclone, together with pPvuM1.9, was used to determine the positions and orientations of the Pvu II genes.

Three types of derivatives were made. First, subclones and a fragment inversion were derived from pPvuM3.4. Second, the insert found in pPvuM1.9 was progressively digested ("resected") from one end with the nuclease Bal31. Third, a number of spontaneous M mutants were isolated by selecting for rare ampicillin-resistant (Amp^r) transformants of E. coli JM107 with pPvuM1.9. Strain JM107 is poorly transformed by this plasmid (as described below), and many of the transformants isolated were found to have copies of pPvuM1.9 that no longer caused production of Pvu II methylase, and had acquired a copy of the 768 bp insertion sequence IS1 [8, 16]. These derivatives are shown in Figure 2, underneath a restriction map of the insert from pPvuM3.4.

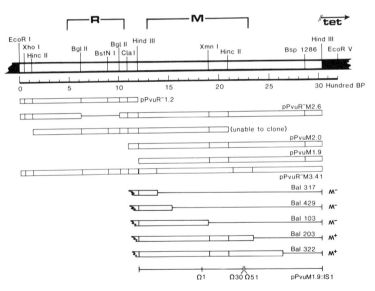

Figure 2. Restriction map and derivatives of inserts from pPvuM1.9 and pPvuRM3.4. The insert from pPvuRM3.4 is shown at the top, along with the inferred gene boundaries for the endonuclease (R), methylase (M), and pBR322-derived tet genes. The derivatives are described in the text. The IS1 insertions shown at the bottom represent three independent mutants of pPvu1.9. Also, note that pPvu⁻M3.41 contains an inversion of the Hind III-Hind III segment.

Each derivative shown in Figure 2 was tested for production, in strain HB101, of endonuclease and methylase activities and for the ability to restrict the growth of bacteriophage *vir*. Each plasmid was also used to program an *E. coli*-derived *in vitro* transcription-translation system (Amersham Corp., Arlington Heights, IL), in the presence of L-[^{35}S] methionine, and the products were sized on SDS-polyacrylamide gels. The results are consistent with the gene boundaries shown at the top of Figure 2.

We believe that the *Pvu* II genes are transcribed divergently from the region between them, as has been found for the *Eco*R V [17] and *Pst* I [18] RMS2s, for the following reasons. First, the *in vitro* translation products associated with the *Pvu* II methylase were full-sized from the Bal203 and Bal322 derivatives, but apparently generated fusion proteins with the pBR322-derived *tet* gene in the other derivatives [8]. Second, the truncated products of pPvuM1.9::IS1Ω1 were considerably smaller than those of Ω30 or Ω51. Third, the product associated with the endonuclease gene was not altered by predigesting pPvuRM3.4 with either *Cla* I or *Hin*d III, though inversion of the 1.9 kb *Hin*d III-*Hin*d III segment (pPvuR⁻M3.41) greatly decreased *in vivo* production of *Pvu* II endonuclease activity. If the carboxy-terminal coding region of the endonuclease gene spanned the *Hin*d III site then digestion with *Cla* I should have removed 20-30 codons, but no reduction in the *in vitro* product size was seen. This data seems most easily explained by suggesting that the entire coding region for the endonuclease lies to the left of *Cla* I, and that the major promoter for the endonuclease gene lies to the right of the *Hin*d III site. This would also be consistent with our ability to isolate pPvuR⁻1.2, and our inability to isolate the Hinc II-*Hin*c II segment as a subclone (Figure 2).

C. Characterization of the *Pvu* II Gene Products

Coelectrophoresis of the *in vitro* products of the *Pvu* II genes with molecular weight markers resulted in size estimates of about 19 kDa for the endonuclease and 34 kDa for the methylase. Compared to other characterized RMS2 proteins these are small, but close to the established size range. For example, the *Bam*H I endonuclease has a subunit molecular weight 22 kDa [19], while the *Dpn* II methylase is 33 kDa [20].

The *Pvu* II methylase may have two distinct forms. *In vitro*, the methylase gene appears to produce a pair of proteins, 34.8 and 33.5 kDa, with the smaller one predominating [8]. Both forms were truncated by Bal31 resection or IS1 insertion, suggesting that the 1.3 kDa size difference arises at the amino end of the methylase. This difference could be due either to protein processing (unlikely in our *in vitro* system) or more than one translation initiation site. While we are currently looking to see if these two forms occur when Pvu II methylase is produced in vivo, such a phenomenon has been seen in vivo with three other RMS2 methylases [21-23; T. Gingeras, pers. commun.]. The biochemical or regulatory significance of the two methylase forms, if any, remains to be established. It would be useful to know how many other RMS2 methylases share this property.

This data, together with that of the previous section, allows us to reconstruct a map of the original plasmid pPvuI (Figure 3). The Pvu II genes appear to take up about 30% of its DNA. Aside from permitting replication in the P. vulgaris, the function of the remaining 70% of pPvuI is unknown. One possibility we have tested, the presence of antibiotic resistance genes, gave negative results: E. coli HB101 carrying pPvuRM5.1 or pPvuRM 5.2 has not acquired resistance to any of 10 tested antibiotics (though we have not looked for a β-lactamase because of the presence of the pBR322 amp gene).

We have found that the Pvu II methylase is a cytosine methylase which probably modifies the internal C of its substrate sequence, 5'-CAGCTC-3'. Extracts of strain HB101(pPvuM1.9) were incubated with pBR322 DNA in the presence of [methyl-^{3}H]S-adenosylmethionine, and the purified labeled DNA was digested to nucleosides enzymatically. Over 95% of the total radioactivity comigrated with 5-methyl-deoxycytidine in high-pressure liquid chromatography on a C18 RPC column developed with a methanol gradient in 0.05M KH$_2$PO$_4$ [R. Blumenthal and H. Schut, unpublished results]. This is consistent with the result of Dobritsa and Dobritsa [24] that cytosine methylation interferes with Pvu II endonuclease action; and of Brooks and Roberts [25] that Pvu II endonuclease cannot digest Alu I-methylated DNA. Alu I methylase modifies the C of its substrate sequence 5'-AGCT-3' [26], and this sequence is contained within the Pvu II substrate sequence. Noncognate methylation within a substrate sequence in some cases interferes with endonuclease action, but in no case has DNA methylation outside a substrate sequence been shown to interfere [25]. Our observation that Alu I endonuclease does not digest a Pvu II-methylated site thus strongly suggests that the Pvu II methylase acts on the internal C [R. Blumenthal, unpublished result]. Our methods cannot rule out N4mC, instead of 5mC, as the modified base generated by the Pvu II methylase. In a list of N4mC-generating methylases, A. Janulaitis [unpubl. data] includes the Pvu II methylase.

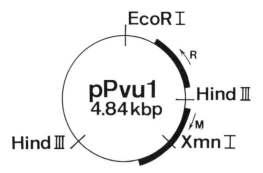

Figure 3. Map of *Proteus vulgaris* plasmid pPvu1. The inferred locations, sizes, and orientations of the *Pvu* II endonuclease (R) and methylase (M) gene are shown. kbp - kilobase pairs.

III. METHYL CYTOSINE RESTRICTION IN *E. COLI*

Two decades have passed since Revel first isolated mutations in two *E. coli* genes, *rgl*A and *rgl*B, that share responsibility for the restriction of glucoseless DNA from T-even bacteriophage [27]. This restriction can be mediated, in the case of bacteriophage T4, by either *rgl*A or *rgl*B alone, and is completely dependent on the presence of 5-hydroxymethyl cytosine (HMC) in the phage DNA [reviewed in ref. 28]. In the case of glucoseless phage T6 mutants, the status of the *rgl*B gene is essentially irrelevant and restriction appears to be mediated by *rgl*A alone [see Table 1, Ref. 28]. Thus the *rgl*A and *rgl*B loci appear to specify restricting activities that have distinct substrate specificities. An endonuclease activity specific for DNA containing HMC, and present only in $rglB^+$ cells, was reported a decade ago [29]. More recently, it has been shown that replacing cytosine with 5 mC in one strand of πX174 or M13 replicative-form DNAs reduces their transfection efficiency by 2- to 100-fold, depending on the degree of replacement [30].

There have thus been scattered hints that *E. coli* strains had some sort of incompatibility with methylcytosine-containing DNA, though it was not clear if these were related phenomena. Furthermore, many *E. coli* strains produce cytosine methylases. For example, *E. coli* K [31] and C strains [32] produce an enzyme that methylates the internal C of the sequence 5'-CC^A_TGG-3' [33]. Recent work with RMS2 methylases, such as the *Pvu* II methylase, have begun to clarify this area of *E. coli* physiology.

A. The Mcr Phenotype

The *Pvu* II plasmids and derivatives shown in Figure 2 transformed *E. coli* HB101 [34] with essentially the same efficiency as the parent vector, pBR322. HB101, however, turned out to be a fortuitous choice as host for our initial cloning. In the course of analyzing the *Pvu* II genes we attempted to move these recombinant plasmids into other host strains, but had very little success. This is shown in Figure 4 which shows the transforming efficiencies of some of the Pvu II plasmids in strain HB101 (open symbols), and in the bacteriophage M13mp vector host strain JM107 [35] (closed symbols and 'x').

Plasmid pBR322 (circles) transformed both strains with roughly equal efficiencies (Figure 4). In contrast pPvuRM3.4 (squares) transformed JM107 at least 1000-fold less efficiently than it transformed HB101. This effect, however, was not due to the Pvu II endonuclease since the same 1000-fold difference was obtained with pPvuM1.9 (triangles). Furthermore, JM107 was efficiently transformed by pPvuM1.9 derivatives such as IS1Ω1 ('x'), implying that poor transformation of JM107 results from an active Pvu II methylase gene.

A number of other E. coli strains have been tested by comparing their transformation efficiencies with pBR322 and pPvuM1.9 (Table II). Of these, 11 were poorly transformed by pPvuM1.9. Only two strain types accepted pPvuM1.9 with an efficiency approaching that of pBR322 (HB101 and RR1 are essentially the same strain; JM107MA2 and ER1389 are derivatives selected for their ability to accept plasmids such as pPvuM1.9 and are discussed below.

TABLE II. Mcr phenotypes of various strains of *Escherichia coli*

Strain	Accepts pPvuM1.9*	Source
HB101	+	Collection
RR1	+	A. Kiss; HB101 $recA^+$
K802	+	H. Revel; C600/HfrC recombinant
JM107MA2	+	JM107 mutant isolated by author
ER1398	+	(Transductant of MM294 isolated and tested by E. Raleigh)
JM107	-	Collection; M13 host
MM294	-	(Tested by E. Raleigh)
JM83	-	Collection; pUC host
B/r	-	Collection
C	-	Collection
W3110	-	Collection
CSR603	-	F. Neidhardt; maxicell strain
C600	-	H. Revel
GM161	-	A. Kiss; *dam*
GM272	-	A. Kiss; *dam dcm*
DH1	-	A. Kiss; *recA endA gyrA*

*"-" means that the transformation efficiency with pPvuM1.9 (carrying the Pvu II methylase gene) was at least 500-fold lower than with the molar equivalent amount of pBR322. Transformants were selected by plating in the presence of ampicillin.

There are several reported examples of DNA methylation affecting gene expression in E. coli [36-41], and it is possible that some DNA methylases could disrupt the expression of essential genes in heterologous hosts. We have found in the case of the Pvu II methylase, however, that DNAs modified by (but not coding for) this enzyme transform strains such as JM107 with reduced efficiency [8]. This result suggests that a form of restriction, active against Pvu II-methylated DNA, is operating in strains such as JM107. Complete Pvu II methylation of pBR322 or bacteriophage λ DNA (1 and 15 Pvu II sites, respectively) led to a roughly 50% reduction in JM107 transforming activity [8]. In contrast, other experiments have indicated a 5×10^5-fold reduction in JM107 transformation by pPvuM1.9, compared to its transformation of HB101. This may reflect the activity of the Pvu II methylase on the host chromosome (roughly 1500 Pvu II sites) leading to "autorestriction."

If this is true, then it should be possible to isolate mutants for the host's "restriction" system that have gained the ability to be transformed by pPvuM1.9. We have, in fact, isolated such mutants using the selection protocol illustrated in Figure 5. With JM107, most of the transformants obtained using a large amount of pPvuM1.9 DNA (10 μg) were methylase-negative and included the IS1 insertion mutants described earlier. But a small number of JM107 Amp^{r}M$^+$ transformants were found and, after being cured of their plasmids with acridine orange and ethidium bromide, were efficiently transformed by pPvuM1.9 [8]. Strain JM107MA2 (methylase accepting) is one of these mutants.

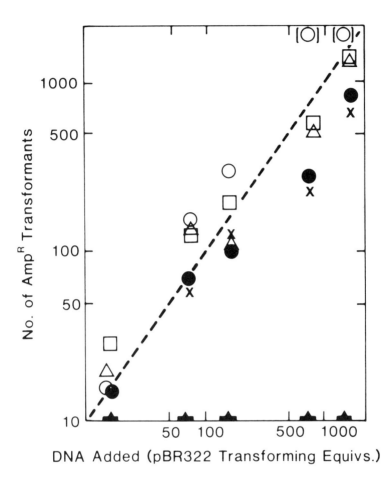

Figure 4. Transformation efficiencies of various plasmids into strains HB101 and JM107, selecting for ampicillin resistance. A stock preparation of pBR322 DNA was tested for its ability to transform strain HB101. One-hundred pBR322 transforming equivalents is the amount of pBR322 DNA that gave an average of 100 amp^r HB101 transformants in several assays. For other plasmids 100 transforming equivalents is the molar equivalent of the amount of pBR322 in 100 pBR322 transforming equivalents. A plasmid that transforms a strain as efficiently as pBR322 transforms strain HB101 will, on average, generate data points on the dotted line. Open symbols refer to transformations of strain HB101; closed symbols and 'x' to transformations of strain JM107. Symbols: O, ●, pBR322 DNA; □, ■, pPvuRM3.4 DNA; △, ▲, PvuM1.9 DNA; x, pPvuM1.9::IS1Ω1 DNA.

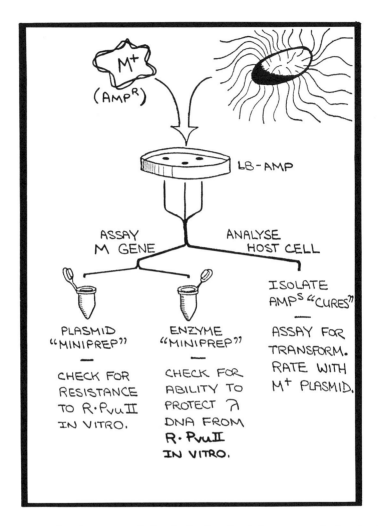

Figure 5. Selection protocol for isolating "methylase-accepting" mutants.

B. Characterization of the Mcr Phenotype

In the course of our work, we learned of others who found that certain cloned RMS2 methylase genes transform most *E. coli* strains poorly. Aside from *Pvu* II [8], the *Bsu*R I [42], SPR (T. Trautner, cited in [42]), and *Hae* II [43] methylase genes were found to exhibit this host strain selectivity. Raleigh and Wilson, in particular, have now studied this phenomenon in some detail and named it the Mcr phenotype (methyl cytosine restriction) [43].

Raleigh et. al. [44] have isolated *E. coli* mutants similar to JM107MA2, and found the mutations to be 90% cotransducible with the *hsd* locus at 99 min. It occurred to them that Revel's *rgl*B locus is close to that location [28], and that *rgl*B specifies an incompatibility with DNA containing modified cytosine. So they looked to see if Mcr is related to *rgl*B [43, 44]. In short, they found that their Mcr⁻ mutants were permissive for glucoseless T4 bacteriophage, and that *rgl*B mutants were efficiently transformed by the cloned *Hae* II methylase gene. We have confirmed this by finding, for several JM107MA mutants, that their efficiency of transformation with pPvuM1.9 appears to vary in parallel with their efficiency of plating glucoseless T4 bacteriophage (Blumenthal and Damschroder, unpublished results). The Mcr locus was named *mcr*B in consonance with *rgl*B, with which it is either tightly linked or identical [43, 44].

To address the question of Mcr sequence-specificity, Raleigh and Wilson made use of their collection of cloned RMS2 methylase genes, and of Mcr⁻ mutants such as those described above (see Table III) [43]. Not surprisingly, the degree of "unacceptability" of some of these methylases varied with their level of expression. Nevertheless, in every case, methylases that poorly transform $mcrB^+$ strains could generate 5'-G*C-3' sequences. Raleigh and Wilson more cautiously suggest R*C as the relevant sequence, since no A*C-generating methylase has yet been tested [43].

The role of Revel's other locus, *rgl*A, is less clear in this context. Raleigh and Wilson have found one methylase gene (*Hpa* II, C *CGG) that transforms $rglA^-$ cells more efficiently than $rglA^+$ cells, but transforms $mcrB^{\pm}$ cells with equal efficiency [43]. They have named *rgl*A, or the *rgl*A-linked gene responsible for this effect, *mcr*A. Since $mcrA^+$ strains can carry the *dcm* methylase, the sequence recognized by *mcr*A cannot be simply 5'-C *C-3'.

C. Implications and Circumvention of the Mcr Phenotype

Why does *E. coli* carry genes such as *mcr*A and *mcr*B? This is unclear, although two possibilities stand out. First, these genes could represent classical restriction systems. If we assume, to make the strongest case, that mcrA recognizes C *CG and that mcrB recognizes G *C, then a fully C-methylated DNA should have a site for one or the other every 13 or so base pairs. A substantial fraction of other cytosine methylases would generate less frequent substrate sites. There are known cases of RMS2 endonucleases active only against a methylated substrate sequence (e.g., Dpn I [45]), but methylated DNAs (e.g., bacteriophage λ grown on HB101 (pPvuRM3.4)) are restricted only 2-200 fold on mcrA$^+$B$^+$ hosts [8, 43]. This is in contrast to 10^5-10_6 fold effects from typical RMS2s. It remains possible that mcrA and/or McrB is more active against 5-hydroxymethyl than 5-methylcytosine, or that the Pvu II substrate sequence represents a suboptimal context for G*C recognition by the mcrB system. It is noteworthy that bacteriophage T4 produces a protein (from the arn gene) that serves to antagonize the activity of the mcrB gene product (see below).

A second possible role for mcrA and/or mcrB is in initiating DNA repair or recombination reactions. This possibility requires that E. coli produce a low-level (damage-specific?) DNA methylase to direct the mcr gene product(s). No such methylase has yet been reported. This system would be comparable to the dam-mutH system of E. coli, in which the mutH endonuclease makes single-strand nicks across from sites modified by the dam methylase [46].

Apart from the possible biological roles of the mcrAB genes, their existence has clear implications for gene cloning work. This is certainly true for cloning from bacteria producing some of the methylase listed in Table III. It could also be true for cloning mammalian DNA into E. coli, since mammalian cells produce a 5'-*CG-3' methylase: roughly a quarter of the methylated sites should be G *CG and may thus be substrates for the mcrB product.

We have examined this possibility using a commercial CG methylase isolated from mouse Krebs II ascites (Amersham Corp., Arlington Heights, IL), pBR322 DNA, and the mcrB$^\pm$ strain pairs (JM107/JM107MA (described above [8, 35], and C600/K802 (K802 came from a cross between C600 and the Cavalli Hfr, and could carry as much as 25 min from the Hfr [47]). The pBR322 DNA was isolated from K802(pBR322) so as to be modified by the EcoK methylase, and was linearized with EcoR I endonuclease as the CG methylase is more active on relaxed DNA substrates [48, 49]. This DNA, 5μg/assay, was incubated with a dilution series of either the CG methylase or Pvu II methylase [pooled peak fractions from a Biogel A0.5m column of an HB101(pPvuM1.9) extract], each in their respective buffer systems. The Pvu II methylase served as a positive control in these experiments. After 3H at 37°C, a mixture including 0.5 U of T4 DNA ligase was added, and after

2 hours at room temperature the DNAs were purified by phenol-chloroform extraction and precipitation. A portion of each DNA preparation was digested with Pvu II endonuclease, or FnuD II or Hha I endonucleases (which are blocked by CG methylation), to assess the degree by a constant factor and used in duplicate transformations, scoring Amp^r transformants. The data, shown in Figure 6, are expressed as the $mcrB^+/mcrB^-$ transforming efficiency ration, and have been normalized to the least-methylated point (at which methylation with either enzyme was undetectable).

TABLE III. Acceptance of cloned methylase genes by Mcr^+ or Mcr^- strains of E. coli [43]

Methylase	Site of Methylation	Accepted by: $mcrB^+$	$mcrB^-$
Adenine Methylase			
EcoR I	GA *ATTC	+	+
Hha II	G *ANTC	+	+
Hind II	GTYR *AC	+	+
Hinf I	G *ANTC	+	+
Pst I	CTGC *AG	+	+
Taq I	TCG *A	+	+
Cytosine Methylases			
BamH I	GGAT *CC	+	+
(dcm	C *C_TGG (A)	+	+)
Alu I	AG *CT	−	+
Dde I	*CTNAG	−	+
Hae III	GG *CC	−	+
Hha I	G *CGC	−	+
Msp I	*CCGG	−	+
Pvu II	CAG *CTG	−	+
Unknown Methylation Site			
Ban I	GGYR CC	−	+
Ban II	GRG CYC	−	+
Bgl I	G CCN₅GGC or GCCN₅GG C	−	+
FnuD II	CGCG	−	+
	CG CG	−	+
Hae II	RG CGCY or RGCG CY	−	+
Hga I	G CGTC and GACG C	−	+
HgiA I	G_TG (A) C_TC (A)	−	+
Nla IV	GGNN CC	−	+

For details, see [43] and text. In all cases specifically tested, the mcrA status was irrelevant to strain compatibility for these methylase genes. "Unaccepted" methylase plasmids transformed cells at least 1000-fold less efficiently than the standard (Mcr^-) strain, E. coli RR1 (HB101 recA). For methylases with unknown methylation sites, the substrate sequence is shown in arrangements consistent with G*C methylation. R, purine; Y, pyrimidine.

The data in Figure 6 indicate that, as expected [8], Pvu II methylation made pBR322 selectively less efficient in transforming JM107 (mcrB$^+$). It appears as if methylation by the mammalian CG methylase also makes pBR322 susceptible to mcrB restriction. This results from a preferential absolute decrease in the transformation of the mcrB$^+$ strain with increased methylation. For example, in the data used to generate Figure 6C, the average transformants of strain K802 decreased from "1" to "128", dropping from 1340 to 1004; but in strain C600 dropped from 460 to 12. The degree of CG methylation in these experiments can only be roughly estimated by the procedure we used: FnuD II and Hha I together should sample 1/8th of the GCG sites (sites that are potential substrates for both the CG methylase and the mcrB activity). The CG-methylated "128" sample appeared to be resistant to cleavage at roughly half of these sites. Nevertheless, it seems clear that DNA modified by the CG methylase is susceptible to mcrB restriction. We have gotten results consistent with this conclusion using DNA from bacteriophage λ, as well (unpublished result).

To summarize, most commonly-used strains of *E. coli* are poorly transformed by certain DNA methylase genes (500-10^5 fold effect), or by DNAs on which those methylases have acted (2-200 fold extract). The poorly transformed strains include several commonly used for genomic cloning (*e.g.*, DH1), and one of the methylases that causes this effect is the mammalian CG methylase (Figure 6). Thus some of the mammalian genomic libraries may have been skewed against more extensively-methylated regions of the chromosome during the initial library amplification.

For initial cloning of mammalian genomic DNA, or DNA from bacteria producing relevant methylase, it would be prudent to use Mcr$^-$ host strains. It may be possible in the future to construct vectors that suppress the mcrB system, using the arn gene of bacteriophage T4. The arn product inhibits the rglB (mcrB) gene produce, possibly through direct interaction [50]. The arn gene has been cloned into pBR325, and the rglB$^+$ strain D0100 efficiently plates glucose T4 bacteriophage if the arn plasmid is present [50].

Much work needs to be done to confirm the identities, substrate specificities, regulation, and biological roles of mcrA and mcrB. Clearly there is as much left to learn regarding the physiology of DNA methylation as there is about the biochemistry of RMS2s such as Pvu II, and work in these two areas often overlaps in unexpected ways.

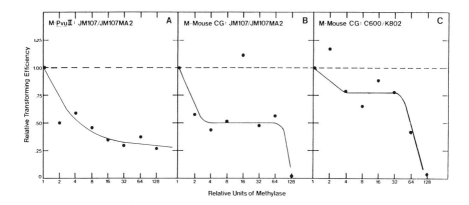

Figure 6. Effects of pBR322 methylation on transformation of $mcr^{\pm}$ strains. Linearized pBR322 was incubated with a dilution series of the Pvu II or mouse CG methylase as described in the text. After 3 hours at 37°C, the reactions were incubated 2 hours more at room temperature with 0.5 U of T4 DNA ligase, then purified as described. The degree of methylation was assessed by digestion with the appropriate restriction endonuclease, and diluted samples were used in duplicate transformation of $CaCl_2$-treated cells, plating on LB agar + 50 µg/ml ampicillin. Each point is the ration of $mcrB^+/mcrB^-$ average Amp^r transformants for a given DNA preparation, which eliminates any differences due to variable DNA recovery or ligation. These have been normalized to the least-methylated point.

A: The "128" sample included 1.25 U of Pvu II methylase (1 U = the amount that protects 1 µg of pBR322 DNA in 1 hour at 37°C from cleavage by Pvu II endonuclease). The "1" sample gave (ave. ± S.E.) 83± transformants of JM107, and 125±26 of JM107MA2.

B: The "128" sample included 8.0 U of mouse CG methylase (1 U = the amount that transfers 1 pmole of S-Ado-Met methyl groups to denatured E. coli DNA in 1 hour at 37°C, as per Amersham). The "1" sample gave 72±0 transformants of JM107, and 144±12 transformants of JM107MA2.

C: The "128" sample included 8.0 U of methylase as in B. The "1" sample gave 460±62 transformants of C600, and 1340±240 transformants of K802.

ACKNOWLEDGEMENTS

The author thanks K. Brennan, J. Cooperider, M. Damschroder, D. Felix, S. Gregory, H. Schut, and T. Samiec for their contributions to various aspects of this work; E. Raleigh and G. Wilson for sharing their unpublished results; J. Brooks, T. Gingeras, and E. Raleigh for their comments on this manuscript; and C. Zimmerman for typing it.

This material is based on work supported by the National Science Foundation under grants PCM-8201953 and DMB-8409652.

REFERENCES
1. D.H. Kruger and T.A. Bickle, Microbiol. Rev. 47, 345-360 (1983).
2. M.G. Marinus in: DNA Methylation: Biochemistry and Biological Significance, A. Razin, H. Cedar, and A.D. Riggs, eds. (Springer-Verlag, New York 1984) pp. 81-109.
3. H.O. Smith and S.V. Kelly in: DNA Methylation: Biochemistry and Biological Significance, A. Razin, H. Cedar, and A.D. Riggs, eds. (Springer-Verlag, New York 1984) pp. 39-71.
4. T.R. Gingeras, L. Greenough, I. Schildkraut, and R.J. Roberts, Nucleic Acids Res. 9, 4525-4536 (1981).
5. R.E. Gelinas, P.A. Myers, and R.J. Roberts, unpublished observations.
6. D.I. Smith, F.R. Blattner, and J. Davies, Nucleic Acids Res. 3, 343-353 (1976).
7. N.L. Brown and M. Smith, FEBS Lett. 65, 284-287 (1976).
8. R.M. Blumenthal, S.A. Gregory, and J.S. Cooperider, J. Bacteriol. 164, 501-109 (1985).
9. J.G. Sutcliffe, Cold Spring Harbor Symp. Quant. Biol. 43, 77-90 (1978).
10. T.A. Bickle, in: Nucleases, S.M. Linn and R.J. Roberts, eds. (Cold Spring Harbor Laboratory, Cold Spring Harbor, NY 1982) pp. 85-108.
11. M. Noyer-Weidner, S. Jentsch, J. Kupsch, M.Bergbauer, and T.A. Trautner, Gene 35, 143-150 (1985).
12. K. Krizsanovich, J. Gen. Virol. 19, 311-320 (1973).
13. J.N. Coetzee, J. Gen. Microbiol. 84, 285-296 (1974).
14. J.N. Coetzee, J. Gen. Microbiol. 87, 173-176 (1975).
15. T.R. Gingeras and J.E. Brooks, Proc. Natl. Acad. Sci. USA 80, 402-406 (1983).
16. H. Ohtsubo and E. Ohtsubo, Proc. Natl. Acad. Sci. USA 75, 615-619 (1978).
17. L. Bougueleret, A. Tsugita, and M. Zabeau, Nucl. Acids Res. 12, 3659-3767 (1984).
18. R.Y. Walder, J.A. Walder, and J.E. Donelson, J. Biol. Chem. 259, 8015-8026 (1984).
19. L.A. Smith and J.G. Chirikjian, J. Biol. CHem. 254, 1003-1006 (1979).
20. B.M. Mannarelli, T.S. Balganesh, B. Greenberg, S.S. Springhorn, and S.A. Lacks, Proc. Natl. Acad. Sci. USA 82, 4468-4472 (1985).
21. U. Gunthert, M. Freund, and T.A. Trautner, J. Biol. Chem. 256, 9340-9351 (1981).
22. U. Gunthert, S. Jentsch, and M. Freund, J. Biol. Chem. 256, 9346-9351 (1981).
23. O.J. Yoo and K.L. Agarwal, J. Biol. Chem. 255, 6445-6449 (1980).
24. A.P. Dobritsa and S.V. Dobritsa, Gene 10, 105-112 (1980).

25. J.E. Brooks and R.J. Roberts, Nucleic Acids Res. 10, 913-934 (1982).
26. V.M. Kramarov and V.V. Smolyaninov, Biokhimiya 46, 1526-1529 (1981).
27. H.R. Revel, Virol. 31, 688-701 (1967).
28. H.R. Revel in: Bacteriophage T4, C.K. Matthews, E.M. Kutter, G. Mosig, and P.B. Berget, eds. (American Society for Microbiology, Washington, DC 1983) pp. 156-165.
29. R.A. Fleischman, J.L. Campbell, and C.C. Richardson, J. Biol. Chem. 251, 1561-1570 (1976).
30. R.Y.-H. Wang, S. Shenoy, and M. Ehrlich, J. Virol. 49, 674-679 (1984).
31. M.G. Marinus, Molec. Gen. Genet. 127, 47-53 (1973).
32. S. Urieli-Shoval, Y. Gruenbaum, and A. Razin, J. Bacteriol. 153, 274-280 (1983).
33. S. Schlagman, S. Hattman, M.S. May, and L. Berger, J. Bacteriol. 126, 990-996 (1976).
34. F. Bolivar and K. Backman, Methods Enzymol. 68, 245-267 (1979).
35. C. Yanisch-Perron, J. Vieira, and J. Messing, Gene 33, 103-119 (1985).
36. S. Hattman, Proc. Natl. Acad. Sci. USA 79, 5518-5521 (1982).
37. R.H.A. Plasterk, M. Vollering, A. Brinkman, and P. Van de Putte, Cell 36, 189-196 (1984).
38. M.G. Marinus, Molec. Gen. Genet. 200, 185-186 (1985).
39. D. Roberts, B.D. Hoopes, W.R. McClure, and N. Kleckner, Cell 43, 117-130 (1985).
40. R.E. Braun and A. Wright, Molec. Gen. Genet. 202, 246-250 (1986).
41. N. Sternberg, J. Bacteriol. 164, 490-493 (1985).
42. A. Kiss, G. Posfai, C.C. Keller, P. Venetianer, and R.J. Roberts, Nucleic Acids Res. 13, 6403-6421 (1985).
43. E.A. Raleigh and G. Wilson, Proc. Natl. Acad. Sci. USA, in press (1976).
44. E.A. Raleigh, R. Trimarchi, and H.R. Revel, in preparation.
45. S. Lacks and B. Greenberg, J. Biol. Chem. 250, 4060-4066 (1975).
46. A.-L. Lu, S. Clark, and P. Modrich, Proc. Natl. Acad. Sci. USA 80, 4639-4643 (1983).
47. W.B. Wood, J. Molec. Biol. 16, 118-133 (1966).
48. Y. Gruenbaum, H. Cedar, and A. Razin, Nature 295, 620-622 (1982).
49. T. Bester and V. Ingram, Prog. Clin. Biol. Res. 198, 95-104 (1985).
50. K. Dharmalingam and E.B. Goldberg, Nature 260, 454-456 (1976).
51. K. Dharmalingam, H.R. Revel, and E.B. Goldberg, J. Bacteriol. 149, 694-699 (1982).

9

Enzymatic Probes for Left-Handed Z-DNA

Franz Wohlrab and Robert D. Wells

Department of Biochemistry
Schools of Medicine and Dentistry
University of Alabama at Birmingham
Birmingham, Alabama 35294

I. INTRODUCTION

DNA is a polymorphic molecule and can adopt a variety of conformations depending mainly on base sequence and environmental conditions [1-4]. Recent interest in this polymorphism has been prompted to a large part by the discovery of Z-DNA, a structure possessing a left-handed helix sense as opposed to the canonical right-handed state (B-DNA) [5,6]. Several findings imply a biological function for Z-DNA or B-to-Z transitions (reviewed in [7,8]).

The structures of a segment of DNA in a right-handed B or in a left-handed Z helix are shown in Figure 1. The structural features have been reviewed in detail previously [2-8]. However, it might be pointed out that virtually all features of the two structures differ, including the sugar conformations, the helical sense, the orientation of some of the glycosidic bonds, the phosphodiester bond torsion angles, etc.

A possible model for the junction between a right-handed B and a left-handed Z helix is shown in Figure 2. Several investigators [reviewed in 2-8] have calculated that the transition can be accomplished within several base pairs (less than 5) and may be effected without unstacking or unpairing of bases.

Left-handed DNA can exist at physiological salt conditions under the influence of supercoiling [9,10 and reviewed in 4,7,8]. The discovery that negative supercoiling induces the formation of the left-handed Z-DNA was pivotal to the studies described in this review, namely the capacity of different enzymes to recognize and utilize left-handed Z-DNA. The vast majority of studies described herein require the use of negative supercoiling to generate the left-handed helices. Furthermore, although the original studies on Z-DNA used alternating (dC-dG) sequences, it has become increasingly clear that perfectly alternating purine-pyrimidine stretches are not a requirement for B-Z transitions [11,12,27-29]. Sequences with the potential to adopt left-handed structures are widespread in nature [13,14] and Z-DNA has been detected in chromosomes from a variety of organisms (reviewed in [4,7]). Also, antibodies to Z-DNA have been found in sera from patients with autoimmune diseases [15] and from autoimmune MRL/l mice [16]. There are a number of ways to detect left-handed DNA, including immunological and chemical methods [45,46].

This article reviews the known interactions of Z-DNA with enzymes and the enzymatic probes available to detect left-handed Z-DNA.

II. NUCLEASES

S1 nuclease from *Aspergillus oryzae*. S1 nuclease recognizes and cleaves structural perturbations at the junctions between contiguous left-

249

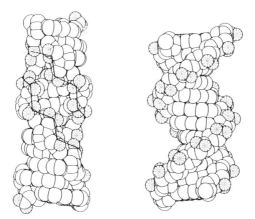

Figure 1. Computer drawn models of left-handed Z-DNA (left side) and right-handed B-DNA (right side). We are grateful to Dr. Stephen C. Harvey (this department) for these figures.

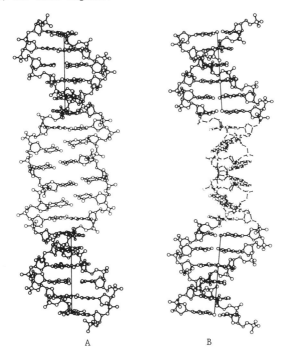

Figure. 2. Models of structural interface between right-handed B and left-handed Z helices. Panels A and B are two mutually perpendicular projections of a minimal patch of Z-DNA embedded in a B-DNA helix. The two central basepairs represent the Z patch. The basepairs immediately above and below are in a transition zone. All other basepairs are regularly helical B segments. This model was published previously [51]; we appreciate the help of Dr. Struther Arnott in providing this figure.

and right-handed DNA. This has been demonstrated for junctions involving left-handed (dC-dG)$_n$ blocks [9,13], (dC-dA)$_n$·(dT-dG) blocks [14] and mixed sequences [15]. S1 nuclease is therefore widely used to probe Z-DNA stretches embedded within right-handed helical regions. The cleavage sites for various junctions map within 3 to 20 base pairs outside of the first alternating residue. More detailed mapping [15,M.J. McLean, personal communication] shows S1 susceptibility over a wider range, with some cleavage sites extending into the left-handed region. Comparison of the extent of cleavage at several junctions of (dC-dG)$_n$ or (dC-dA)$_n$ with various non-alternating regions demonstrated different susceptibilities. In the case of (dT-dG)$_n$·(dC-dA)$_n$ blocks that flank the rat somatostatin gene, the entire left-handed segment is sensitive to attack by S1 nuclease [22]. The B-Z junctions appear to be located within the alternating purine-pyrimidine stretches and the S1 cleavage pattern is strongly dependent on the nature of the adjacent non-Z-DNA sequences. Recent studies on the thermodynamics of B-Z transitions [11,12,23] also indicate conformational heterogeneity of junctions. The flexibility of junctions is therefore sequence- and supercoil dependent [13,24].

The mechanism of the S1 reaction with the B-Z interface is not known. The enzyme was originally isolated as a single-strand specific endonuclease. This observation, together with the finding that $\Delta G_{junction}$ increases with increasing NaCl concentrations [25], and the theoretical unwinding of the adjacent right-handed helix by about 0.4 turns/junction [23,25], would be consistent with some single-strand character of the B-Z junctions. However, chemical studies [24] using bromoacetaldehyde (BAA) have demonstrated that the B-Z interface contains few, if any unpaired bases. Furthermore, the S1 nuclease hypersensitivity of other types of double-stranded sequences [27,29] indicates that single-strandedness is not necessary for cleavage. Thus, the precise structure recognized by S1 nuclease remains to be elucidated.

Other nucleases. The B-Z interface is recognized by a number of nucleases besides S1. Use of these enzymes revealed conformational differences between various junctions [14,30]. Mung bean nuclease catalyzes cleavage of junctions adjacent to left-handed (dC-dG)$_n$ regions but does not cleave within junctions flanking (dC-dA)$_n$·(dT-dG)$_n$ regions [14,30].

Although S1 nuclease is a highly specific enzyme, its major disadvantage as a probe for left-handed DNA lies in the fact that the pH optimum for endonucleolytic cleavage is around 4.6. The BAL31 nuclease has been used at neutral pH values to detect the conformational aberrations within B-Z junctions [30]. This enzyme is active at high NaCl concentrations (several molar) and can therefore be used to investigate the salt-induced B-Z transition. However, the associated potent exonuclease activity limits its use for fine mapping of junction regions at the base pair level. However, the BAL31 nuclease is an excellent probe if long sequences (several kbp) are explored, especially for salt-induced conformational changes.

Nuclease P1 from *Penicillium citrinum* recognizes a number of structural features at neutral pH [31]. Although its pH optimum for double strand cleavage is acid, the enzyme will catalyze site-specific cleavage of junctions between right- and left-handed DNA sequences at neutral pH (J.A. Blaho, personal communication). This result is important since it demonstrates that B-Z junctions (or Z-helices) are not favored by low pHs.

III. RESTRICTION ENDONUCLEASES AND METHYLASES

Hha I and *Bss*H II. Because of the dramatically altered DNA conformation in Z-DNA compared to B-DNA, several groups have investigated the capacity of left-handed DNA to serve as a substrate for modification methylases and restriction endonucleases. Studies on DNA polymers containing alternating G and 5-methyl-C moieties (see [32,33]) and on recombinant plasmids [34] demonstrate the stabilizing effect of methyl groups on Z-DNA. In negatively supercoiled DNA, the *Hha* I modification methylase (M*Hha* I) will not methylate its target site d(GCGC) when it is in a left-handed conformation [35,36]. Likewise, the *Hha* I restriction endonuclease does not catalyze cleavage at left-handed d(GCGC) sites. The *Bss*H II restriction endonuclease recognition sequence is d(GCGCGC). It is not cleaved by the restriction enzyme when it is part of a left-handed structure [35]. In the case of M*Hha* I, a low amount of methylation of $(dC-dG)_n$ tracts is observed in conditions under which these are completely left-handed [36]. This may be due to the B-Z dynamic equilibrium.

In summary, these results show that some features of the Z-helix (the helical sense or the sugar conformation or the glycosidic bond orientations, etc.) are not compatible with cleavage, as expected. However, this observation may be used as a probe for left-handed helices in other genomes [12,36].

Other restriction endonucleases. Structural perturbations within B-Z Junction regions have also been demonstrated using restriction endonucleases. When a *Bam*H I site (GGATCC) neighbors a $(dC-dG)_n$ tract, cleavage by the *Bam*H I restriction enzyme was inhibited by the formation of left-handed DNA in the flanking sequences [13]. In a related study [37], the *Bam*H I recognition sequence was removed various distances from a left-handed segment. Cleavage by *Bam*H I was inhibited at a distance of 4 or less base pairs from a left-handed $(dG-dC)_n$ block, but no inhibition was observed when the site was 8 bp away. This result supports the notion of a maximal length of 8 bp for the junction. However, prior Raman spectroscopic studies [38] revealed that the B-structure was perturbed at a distance of several turns of helix (30-40 bp) away from the Z-helix.

Similar results were found with EcoR I when an EcoR I site (GAATTC) replaced the BamH I site in the work described above (M. Caserta, W.T. Hsieh, and R.D. Wells, unpublished studies).

IV. INTERACTION OF LEFT-HANDED DNA WITH OTHER ENZYMES

RNA polymerase. Attempts to utilize DNA sequences in left-handed conformations as templates in vitro and in vivo for RNA polymerases revealed, in general, reduced transcription [39-42]. For example, transcription by E. coli RNA polymerase is almost totally blocked by a (dC-dG)$_{16}$ sequence in left-handed conformation in vitro [43], whereas (dC-dA)$_{21}$·(dT-dG)$_{21}$ in left-handed form has little or no effect on transcription. It therefore appears that not all left-handed sequences are effective transcriptional blocks. In addition, RNA polymerase can transcribe plasmids in vivo containing the (dG-dC)$_{16}$ block, indicating that either the superhelical density within the cell is not high enough to cause a B-Z transition or that additional factors facilitate the readthrough of the left-handed sequences.

rec1 protein from Ustilago. The rec1 protein from Ustilago catalyzes homologous pairing of DNA molecules. It binds Z-DNA 75 times stronger than B-DNA in an ATP-dependent manner [44] and DNA pairing is strongly promoted by sequences of Z-DNA [45]. The pairing reaction appears to start within the Z-DNA stretch and can be blocked by addition of anti-Z-DNA antibody. Left-handedness, however, is not sufficient for linking by rec1, but has to be accompanied by homology of two DNA stretches. These results suggest that Z-DNA is an intermediate in the rec1 catalyzed homologous DNA pairing reaction. Recent studies with the E. coli recA protein, which is generally believed to be functionally identical to the Ustilago rec1 protein, reveal similar results qualitatively but not quantitatively (J.A. Blaho and R.D. Wells, unpublished).

Repair enzymes. B and Z DNA are equally methylated by dimethyl sulfate. Under alkaline conditions, the imidazole ring of 7-methyl-guanine residues reacts to yield the ring-opened form 2,6-diamino-4-oxo-5-methylformamidopyrimidine. This modified nucleotide is an inhibitor of DNA replication. The lesion can be removed enzymatically by a DNA glycosylase from modified poly(dG-dC) in the B conformation but not in the Z form [46]. It has also been reported that the activity of the E. coli O^6-methyl guanine-DNA methyl transferase is lower on Z-DNA than on B-DNA [47,48].

V. SUMMARY

In conclusion, one of the aspects of the DNA polymorphism observed is the formation of Z-DNA under a variety of conditions. Left-handed DNA stretches not only represent alternate structures, but also exert long-range effects due to their influence on superhelical properties on an entire supercoiled DNA as first shown six years ago [49,50]. The examples given in this review emphasize the site-specificity of enzymes due to structural features rather than sequence itself. In this fashion, the reversible transition from B to Z DNA could modulate site-specific events on many levels of biological regulation.

Considering all of the enzymes studied to date (S1, mung bean, BAL31, P_1 nucleases, Hha I, BssH II, MHha I, BamH I, EcoR I, RNA polymerase, recl, recA, DNA glycosylase, O^6-methylguanine-DNA methyltransferase), only the recl (and possibly the recA) protein seems to recognize and utilize left-handed DNA.

A large number of questions related to the biology of Z-DNA are unanswered including: what is the DNA structure (B or Z or other) which is in physical contact with proteins; is Z-DNA recognized by proteins or are junctions the important features; do proteins revert the Z structure to B or to some other right-handed conformation; what other cofactors (perhaps chiral in nature) may be involved; what are the alternate forms of left-handed DNA; does left-handed DNA exist *in vivo*; what is the biological role(s) of left-handed DNA? The future of this field of investigation will be exciting indeed.

VI. ACKNOWLEDGEMENTS

This work was supported by grants from the N.I.H. (GM 30822) and the N.S.F. (83-08644).

VI. REFERENCES

1. R.D. Wells, T.C. Goodman, W. Hillen, G.T. Horn, R.D. Klein, J.E. Larson, U.R. Muller, S.K. Neuendorf, N. Panayotatos, and S.M. Stirdivant, Prog. Nucleic Acids Res. Mol. Biol. 24, 167-267 (1980).
2. S.B. Zimmerman, Ann. Rev. Biochem. 51, 395-427 (1982).
3. R.E. Dickerson, H.R. Drew, B.N. Connor, R.M. Wing, A.V. Fratini, and M.L. Kopka, Science 216, 475-478 (1982).
4. C.K. Singleton in: Cell Proliferation: Recent Advances (Leffert, H.L. and Boynton, A.L., eds. 1984), Academic Press, New York.
5. A.H. -J. Wang, G.J. Quigley, F.J. Kolpak, J.L. Crawford, J.H. van Boom, G. van der Marel, and A. Rich, Nature 282, 680-686 (1979).

6. H.R. Drew, T. Takano, K. Itakura, and R.E. Dickerson, Nature 286, 567-573 (1980).
7. A. Rich, A. Nordheim, and A.H.-J. Wang, Ann. Rev. Biochem. 53, 791-846 (1984).
8. R.D. Wells, B.F. Erlanger, H.B. Gray, L.H. Hanau, T.M. Jovin, M.W. Kilpatrick, J. Klysik, J.E. Larson, J.C. Martin, J.J. Miglietta, C.K. Singleton, S.M. Stirdivant, C.M. Veneziale, R.M. Wartell, C.F. Wei, W. Zacharias, and D. Zarling, Gene Expression, UCLA Symp. Mol. Cell. Biol. 8, 3-18 (1983).
9. C.K. Singleton, J. Klysik, S.M. Stirdivant, and R.D. Wells, Nature 299, 312-316 (1982).
10. L.J. Peck, A. Nordheim, A. Rich, and J.C. Wang, Proc. Natl. Acad. Sci. USA 79, 4560-4564 (1982).
11. M.J. Ellison, R.J. Kelleher, III, A.H.-J. Wang, J.F. Habener, and A. Rich, Proc. Natl. Acad. Sci. USA 82, 8320-8324 (1985).
12. M. McLean, J.A. Blaho, M.W. Kilpatrick, and R.D. Wells, Proc. Natl. Acad. Sci. USA 83, in the press (1986).
13. C.K. Singleton, J. Klysik, and R.D. Wells, Proc. Natl. Acad. Sci. USA 80, 2447-2451 (1983).
14. C.K. Singleton, M.W. Kilpatrick, and R.D. Wells, J. Biol. Chem. 259, 1963-1967 (1984).
15. M.W. Kilpatrick, J. Klysik, C.K. Singleton, D. Zarling, T.M. Jovin, L.H. Hanau, B.F. Erlanger, and R.D. Wells, J. Biol. Chem. 259, 7268-7274 (1984).
16. H. Hamada, M.G. Petrino, T. Kakunaga, and A. Novel, Proc. Natl. Acad. Sci. USA 79, 6465 (1982).
17. R. Thomas, S. Beck, and F.M. Pohl, Proc. Natl. Acad. Sci. USA 80, 5550-5553 (1983).
18. E.M. Lafer, R.P.C. Valle, A. Moller, A. Nordheim. P.H. Schur, A. Rich, and B.D. Stollar, J. Clin. Invest. 71, 314-321 (1983).
19. H.R. Bergen, III, M.J. Losman, T.R. O'Connor, W. Zacharias, J.E. Larson, R.D. Wells, and W.J. Koopman, J. Immun., submitted 1986.
20. B.H. Johnston and A. Rich, Cell 42, 713-724 (1986).
21. G. Galazka, E. Palacek, R.D. Wells, and J. Klysik, J. Biol. Chem. (in press, 1986).
22. T.E. Hayes and J.E. Dixon, J. Biol. Chem. 260, 8145-8156 (1985).
23. T.R. O'Connor, D.S. Kang, and R.D. Wells, J. Biol. Chem., submitted 1986.
24. D.S. Kang and R.D. Wells, J. Biol. Chem. 260, 7783-7790 (1985).
25. F. Azorin, A. Nordheim, and A. Rich, EMBO J. 2, 649-655 (1983).

26. L.J. Peck and J.C. Wang, Proc. Natl. Acad. Sci. USA 80, 6206-6210 (1983).
27. E. Schon, T. Evans, J. Welsh, and A. Efstratiadis, Cell 35, 837-848 (1983).
28. D.E. Pulleyblank, D.B. Haniford, and A.R. Morgan, Cell 42, 271-280 (1985).
29. C.R. Cantor and A. Efstratiadis, Nucleic Acids Res. 21, 8059-8072 (1984).
30. M.W. Kilpatrick, C.-F. Wei, H.B. Gray, and R.D. Wells, Nucleic Acids Res. 11, 3811-3818 (1983).
31. D.B. Haniford and D.E. Pulleyblank, Nucleic Acids Res. 13, 4343-4363 (1985).
32. M. Behe and G. Felsenfeld, Proc. Natl. Acad. Sci. USA 78, 1619-1623 (1981).
33. J. Feigon, A.H.-J. Wang, G.A. van der Marel, J.H. van Boom, and A. Rich, Nucleic Acids Res. 12, 1243-1263 (1984).
34. J. Klysik, S.M. Stirdivant, C.K. Singleton, W. Zacharias, and R.D. Wells, J. Mol. Biol. 168, 51-71 (1983).
35. L. Vardimon and A. Rich, Proc. Natl. Acad. Sci. USA 81, 3268-3272 (1984).
36. W. Zacharias, J.E. Larson, M.W. Kilpatrick and R.D. Wells, Nucleic Acids Res. 12, 7677-7692 (1984).
37. F. Azorin, R. Hahn, and A. Rich, Proc. Natl. Acad. Sci. USA 81, 3268-3272 (1984).
38. R.M. Wartell, J. Klysik, W. Hillen, and R.D. Wells, Proc. Natl. Acad. Sci. USA 79, 2549-2553 (1982).
39. J.H. van de Sande and T.M. Jovin, EMBO Journal 1, 115-120 (1982).
40. R. Durand, C. Job, D.A. Zarling, M.Teissere, T.M. Jovin and D. Job, EMBO Journal 2, 1707-1714 (1983).
41. J.J. Butzow, Y.A. Shin, and G.L. Eichhorn, Biochemistry 23, 4837-4843 (1984).
42. C. Santoro, F. Costanzo, and G. Ciliberto, EMBO J. 3, 1553-1559 (1984).
43. L.J. Peck and J.C. Wang, Cell 40, 129-137 (1985).
44. E.B. Kmiec, K.J. Angelides, and W.K. Holloman, Cell 40, 139-145 (1985).
45. E.B. Kmiec and W.K. Holloman, Cell 44, 545-554 (1986).
46. C. Lagravere, B. Malfoy, M. Leng, and J. Lavel, Nature 310, 798-800 (1984).
47. M. Richirawat, F.F. Becker, and J.N. Lapeyre, Nucleic Acids Res. 12, 3357-3372 (1984).
48. S. Boiteux, R. Costa de Oliveira, and J. Laval, J. Biol. Chem. 260, 8711-8715 (1985).

49. J. Klysik, S.M. Stirdivant, J.E. Larson, P.A. Hart, and R.D. Wells, Nature 290, 672-677 (1981).
50. S.M. Stirdivant, J. Klysik, and R.D. Wells, J. Biol. Chem., 257, 10159-10165 (1982).
51. S. Arnott, R. Chandrasekaran, I.H. Hall, L.C. Puigjaner, J.K. Walker, and M. Wang, Cold Spring Harbor Symposium 47, 53-65 (1983).

10

Enhancement of the Apparent Cleavage Specificities of Restriction Endonucleases: Applications to Megabase Mapping of Chromosomes

Michael McClelland* and Michael Nelson**

*Department of Biochemistry and Molecular Biology
University of Chicago
920 East 58th Street
Chicago, IL 60637

**New England Biolabs, Inc.
32 Tozer Road
Beverly, MA 01915

I. INTRODUCTION

Restriction endonucleases generally recognize DNA sequences of four to six base pairs [1]. Therefore, these enzymes cut DNA that contains equal amounts of A,C,G and T, distributed at random, into fragments with an average size of 4^4 to 4^6 base pairs, (256 to 4096 base pairs). Accordingly, restriction endonuclease mapping techniques are appropriate for the analysis of gene-sized DNA molecules. For example, a restriction endonuclease with a six base specificity will cut the human genome into approximately 750,000 fragments averaging 4 kilobase pairs. However, more specific DNA cutting methods would yield fewer and larger fragments, suitable for the analysis of large, complex genomes.

Techniques have been described for the preparation and separation of DNA molecules of one megabase pair or more [2,3,4,5]. To produce fragments of this size from random DNA one must have a recognition sequence which is the log (base 4) of 1,000,000; approximately ten base pairs. The genome sizes of various organisms and the number of fragments generated by cleavage systems of varying specificities are presented in Table 1.

We describe here a number of strategies which can be used to increase the apparent specificity of restriction endonucleases to produce Average Fragment Sizes (AFS) in the range of 6,000 to 270,000,000 base pairs. These are:
1. Protection of restriction endonuclease recognition sites from cleavage by sequence-specific DNA methylation.
2. Generation of cleavage sites using sequence-specific DNA methylation and a methylation-dependent restriction endonuclease, Dpn I.
3. Blocking DNA methylases by other methylases.
4. Prediction of rare restriction recognition sites for particular genomes based on the non-random sequence arrangement of natural DNAs.

We believe that combinations of the above-mentioned three methods will provide a framework for the eventual molecular dissection of large, complex, DNA molecules.

II. CROSS-PROTECTION OF RESTRICTION ENDONUCLEASES BY SEQUENCE-SPECIFIC METHYLASES

Type II restriction endonucleases are named for their ability to restrict the entry of bacteriophage into bacterial cells [7]. Invading phage DNA is cut enzymatically at four to six base recognition sequences*, provided that these sequences are not methylated. Methylation takes place at the N-6 position of adenine or the C-5 or N-4 position of cytosine within

---
*Other specific endonucleases with longer recognition sequences include Rsr II (CGG(A/T)CCG), Aci I (CCAPyN$_5$TGC), Sfi I (GGCCN$_5$GGCC), and Not I (GCGGCCGC). These might not be restriction endonucleases because their recognition sequences occur so rarely in DNA. One possible role is in sequence specific recombination.

TABLE 1. Chromosome complexity ladder. The columns are: A log scale of DNA size in base pairs; Effective recognition sequence size (ERSS) of a cleavage system that generates an average fragment size (AFS) specified in the first column, e.g. 10^6 base pairs is the AFS of a 10.0 base pair ERSS; Genome sizes for some representative species [6]; The number of fragments generated by an eight, ten and twelve base pair ERSS on a genome of the size specified in the first column.

Base Pairs	ERSS	Species	Number of Fragments with an ERSS of		
			8	10	12
10^6	16.6	Wheat	153,000	9,540	600
10^9	14.9	Mammals	15,300	954	60
10^8	13.3	Drosophila Arabidopsis	1,530	95	6
10^7	11.5	Yeast	153	10	0
10^6	10.0	E. coli	15	1	0
10^5	8.3	Vaccinia virus T4 phage	2	0	0
10^4	6.4	Lambda phage ϕX174, SV40	0	0	0

the recognition sequence [8]. Restriction endonucleases do not recognize DNA methylated by the corresponding modification methylase at this 'cognate' recognition site.

Furthermore, many restriction endonucleases are sensitive to methylation at other 'non-cognate' sites within their recognition sequences, produced in vitro by other DNA methylases [9]. For instance, Hind III does not cut ^{m}AAGCTT methylated by M.Hind III, (cognate methylation) or AAG^{m}CTT methylated by M.Alu I (AG^{m}CT) (non-cognate methylation) [9]. This inability of restriction endonucleases to cut methylated DNA can be utilized to protect a subset of restriction recognition sequences by sequence-specific methylation [10,11,12]. For example, the sequence-specific DNA methylase dam (G^{m}ATC) protects against Cla I (ATCGAT) cleavage at G^{m}ATCGAT or its complement, ATCG^{m}ATC. Consequently, the specificity of Cla I after dam methylation of DNA is (A/T/C)ATCGAT(A/T/G), equivalent to a 6.44 pair recognition sequence (AFS 7512 base pairs).

The more overlap there is between a restriction recognition sequence and a DNA methylase recognition sequence, the more specific the effective restriction specificity becomes after methylation. DNA methylases that have recognition sequences of fewer than four bases are of particular interest. One such methylase, which we have purified (Nelson, unpublished), is M.H I (GG^{m}CC and G^{m}CNGC). Several methylases with short recognition sequences (two to three base pairs) have been described (e.g. 26-32). In principle, such enzymes could increase the apparent specificity of many restriction endonucleases because of the large number of potential overlaps. However, these methylases are generally not well characterized.

One way of further increasing the apparent specificity of a restriction endonuclease is to use more than one methylase before cleavage. Using Cla I as an example, if one methylates with both dam (G^{m}ATC) and M.Hha II (G^{m}ANTC), Cla I is blocked at G^{m}ATCGA, ATCG^{m}ATC, GA^{m}ATCGAT and ATCG^{m}ATTC. This is equivalent to a 6.51 base pair recognition sequence (1-(3/4 x 3/4 x 15/16 x 15/16) +6), (AFS 8256 base pairs). A list of known methylase cross-protection enhancements of restriction specificity that yield Effective Recognition Sequence Size (ERSS) of over six base pairs are given in Table 2. A list of 200 potential enhancements has been compiled (not shown).

Other methylation specificities of less than four base pairs can potentially be achieved by performing methylation reactions under "relaxed" conditions, such as with 30-50% glycerol or alkaline pH (see [9] for a review). In at least one case a mutant methylase has been isolated that has a broader specificity than wild type [14].

In general, we expect hexamer methylation cross-protection schemes to give 6.4 to 7 base pairs cleavage specificities, i.e. DNA fragments ranging

TABLE 2. High specificity DNA cleavage using methylase/endonuclease cross-protection. Some methylase/endonuclease cross-protections are as follows: M.H I and M.*Hha* II (Nelson, unpublished), M.*Hin*f III [13], and M.*Nla* IV (R. Camp, unpublished).

Methylase	Methylase Specificity	Restriction Endonuclease	Restriction Specificity	New Cleavage Specificity	ERSS
M.<u>Alu</u> I	AGmCT	<u>Nhe</u> I	GCTAGC	C A GGCTAGC T T	6.44
M.<u>dam</u>	GmATC	<u>Cla</u> I	ATCGAT	C G ATCGATA T T	6.44
M.<u>dam</u> + *M.<u>Hha</u> II	GmATC GmANTC	<u>Cla</u> I	ATCGAT	-----	6.51
*M.<u>H</u> I	GCNGC + GGmCC	<u>Pst</u> I	CTGCAG	C G ACTGCAGA T T	6.44
*M.<u>H</u> I	GCNGC + GGmCC	<u>Pvu</u> II	CAGCTG	C G ACAGCTGA T T	6.44
M.<u>Hae</u> III	GGmCC	<u>Bgl</u> I	GCCN$_5$GGC	C G AGCCN$_5$GGCA T T	6.44
M.<u>Hae</u> III	GGmCC	<u>Nae</u> I	GCCGGC	C G AGCCGGCA T T	6.44
*M.<u>Hin</u>f III	CGAmAT	<u>EcoR</u> I	GAATTC	G C AGAATTCA T T	6.44
*M.<u>Nla</u> IV	GGNNmCC	<u>Apa</u> I	GGGCCC	C G AGGGCCCA T T	6.44

from 7,000 to 17,000 base pairs. Such fragment sizes are ideal for dissecting the genomes of moderate to large DNA viruses. In combination with methods described in Section III, fragment sizes in the 100,000 base pair range may be predicted for certain DNAs.

III. GENERATION OF CLEAVAGE SITES FOR A METHYLATION DEPENDENT RESTRICTION ENDONUCLEASE

8 to 14 base pair recognition sequences (AFS 65,000 to 270,000,000 base pairs) are suitable for the analysis of chromosome-sized DNA molecules.

A method which allows such highly selective cleavage employs Dpn I, which cleaves DNA only when both strands have been methylated at the sequence G^mATC [15,16]. Since most DNAs such as those of mammals, insects and plants are not methylated at G^mATC, these DNAs are not cut by Dpn I. However, this methylated sequence can be created in vitro using certain sequence-specific adenine methylases whose specificities overlap the Dpn I recognition sequence. For instance, M.Taq I (TCG^mA) methylation enables Dpn I to cleave at the eight base pair sequences TCGATCGA [17]. Similarly, M.Mbo II ($GAAG^mA$) methylation results in new Dpn I cleavage sites [18] at the inverted repeat:

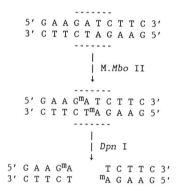

Furthermore, M.Mbo II can be used in conjunction with other methylases to produce cleavage by Dpn I at other rare sequences. For instance, the use of M.Mbo II and M.Cla I together produces cleavage of DNA not only at the M.Mbo II/M.Mbo II inverted repeat GAAGATCTTC and the M.Cla I/M.Cla I overlapping repeat ATCGATCGAT, but also at hybrid sites containing both M.Mbo II and M.Cla I recognition sequences; GAAGATCGAT and its complement ATCGATCTTC. These four ten base recognition sequences give the M.Mbo II/M.Cla I/Dpn I site a frequency equivalent to a nine base recognition sequence. Such a sequence should occur on average once in 262,000 base pairs of random DNA sequence.

However, methylases may be inhibited by methylation at non-cognate sites. Thus, for instance, some DNA methylases such as M.*Cla* I (ATCG^{m}AT) may not act on recognition sequences containing ^{m}CpG at AT^{m}CGAT. This complication would limit the use of this particular enzyme in the methylase/*Dpn* I cleavage of mammalian DNA, which is methylated at ^{m}CpGs. Such possibilities are being investigated.

A number of highly selective DNA cutting tools using sequence-specific adenine methylation and *Dpn* I cleavage have been demonstrated. Others may be predicted based on the known and potential specificities of modification methylases. All the known and some of the potential cleavage systems derived from known restriction-modification systems are listed in Table 3. These cleavage systems potentially generate fragments averaging from 50,000 to 270,000,000 base pairs. When combined with the ability to separate the large fragments generated by pulsed field gel electrophoresis [2,3] these cutting strategies should be useful for generating physical maps of very large DNA molecules and for isolating large fragments from chromosomes.

For example, the *E. coli* genome is 4 x 10^6 base pairs [6]. The ten base pair recognition sequence for M.*Cla*/*Dpn* I, (ATCGATCGAT), should occur in this chromosome approximately 4 times. The M.*Bsp*H I/*Dpn* I (TCATGATCATGA) recognition sequence should occur 8 times in an average human chromosome of 150 x 10^6 base pairs. Therefore, methylase/*Dpn* I cleavage strategies are suitable for linkage mapping over megabases of chromosomal DNA, chromosome walking and the production of subchromosomal libraries [17,19].

Methylation-dependent restriction endonucleases other than *Dpn* I will also generate rare cleavage systems with overlapping methylases. For instance, a screen of 30 species has produced one cytosine methylation dependent endonuclease from *Flavobacterium* with a four or five base specificity (Morgan and Schildkraut, unpublished). This endonuclease should yield highly specific cleavage schemes in conjunction with appropriate cytosine methylases.

Many other methylation-dependent endonucleases can be expected to be identified.

IV. BLOCKING OF METHYLATION BY PRIOR SEQUENCE-SPECIFIC DNA METHYLATION

A strategy to enhance restriction endonuclease cleavage specificity by altering the specificity of the corresponding modification methylase is outlined in this section (McClelland and Nelson, submitted).

The methylase M.*Bam*H I, methylates at the first cytosine in the DNA sequence GGATCC to produce GGAT^{m}CC. M.*Bam*H I is inhibited by prior methylation at the second cytosine; GGATC^{m}C. In contrast, cleavage of DNA by the endonuclease *Bam*H I is not inhibited by GGATC^{m}C but is inhibited by GGAT^{m}CC.

TABLE 3. Methylase/*Dpn* I recognition sequences. Available and potential cleavage systems using *Dpn* I and either one or two site specific methylases are presented. More complicated schemes employing more than two methylases can be envisioned but are not included here.
AFS = Average fragment size; ERSS = Effective recognition sequence site

Methylases	Recognition Sequence[#]	ERSS[@]	AFS
M.*Acc* III	TCCG<u>GATC</u>CGGA	12.0	16,800,000
M.*Acc* III/M.*Eco*D XI	TCCG<u>GATC</u>AN$_7$ATTC	11.7	11,000,000
M.*Acc* III/M.*Eco*B	TCCG<u>GATC</u>AN$_5$TGCT	11.7	11,000,000
M.*Acc* III/M.*Cla* I	TCCG<u>GATC</u>GAT	9.7	700,000
M.*Acc* III/M.*Mbo* II	TCCG<u>GATC</u>TTC	9.7	700,000
M.*Acc* III/M.*Mme* I	TCCG<u>GATC</u>CRAC	10.4	1,900,000
M.*Acc* III/M.*Nru* I	TCCG<u>GATC</u>GCGA	11.0	4,200,000
M.*Acc* III/M.*Bsp*H I	TCCG<u>GATC</u>ATGA	11.0	4,200,000
M.*Acc* III/M.*Taq* I	TCCG<u>GATC</u>GA	7.9	57,000
M.*Acc* III/M.*Taq* II	TCCG<u>GATC</u>GGTC	11.0	4,200,000
(M.*Acc* III/M.*Xba* I)	TCCG<u>GATC</u>TAGA	11.0	4,200,000
M.*Bsp*H I	TCAT<u>GATC</u>TAGA	12.0	16,800,000
M.*Bsp*H I/M.*Eco*B	TCAT<u>GATC</u>AN$_7$ATTC	11.7	11,000,000
M.*Bsp*H I/M.*Eco*D XI	TCAT<u>GATC</u>AN$_5$TGCT	11.7	11,000,000
M.*Bsp*H I/M.*Cla* I	TCAT<u>GATC</u>GAT	9.7	700,000
M.*Bsp*H I/M.*Mbo* II	TCAT<u>GATC</u>TTC	9.7	700,000
M.*Bsp*H I/M.*Mme* I	TCAT<u>GATC</u>CRAC	10.4	1,900,000
M.*Bsp*H I/M.*Nru* I	TCAT<u>GATC</u>GCGA	11.0	4,200,000
M.*Bsp*H I/M.*Taq* I	TCAT<u>GATC</u>GA	7.9	57,000
M.*Bsp*H I/M.*Taq* II	TCAT<u>GATC</u>GGTC	9.7	700,000
M.*Bsp*H I/M.*Xba* I	TCAT<u>GATC</u>TAGA	11.0	4,200,000
M.*Eco*B	AGCAN$_5$T<u>GATC</u>AN$_5$TGCT	14.0	270,000,000
M.*Eco*B/M.*Eco*D XI	AGCAN$_5$T<u>GATC</u>AN$_7$ATTC	13.0	67,000,000
M.*Eco*B/M.*Cla* I	AGCAN$_5$T<u>GATC</u>GAT	9.9	900,000
M.*Eco*B/M.*Mbo* II	AGCAN$_5$T<u>GATC</u>TTC	9.9	900,000
M.*Eco*B/M.*Mme* II	TCAT<u>GATC</u>CRAC	10.4	1,900,000
M.*Eco*B/M.*Nru* I	AGCAN$_5$T<u>GATC</u>GCGA	11.7	11,000,000
M.*Eco*B/M.*Taq* I	AGCAN$_5$T<u>GATC</u>GA	7.9	57,000
M.*Eco*B/M.*Taq* II	AGCAN$_5$T<u>GATC</u>GGTC	11.7	11,000,000
(M.*Eco*B/M.*Xba* I)	AGCAN$_5$T<u>GATC</u>TAGA	11.7	11,000,000
M.*Eco*D XI	GAATN$_7$T<u>GATC</u>AN$_7$ATTC	14.0	270,000,000
M.*Eco*D XI/M.*Cla* I	GAATN$_7$T<u>GATC</u>GAT	9.9	900,000
M.*Eco*D XI/M.*Mbo* II	GAATN$_7$T<u>GATC</u>TTC	9.9	900,000
M.*Eco*D XI/M.*Mme* I	GACAN$_5$TGATCCRAC	10.9	4,000,000
M.*Eco*D XI/M.*Nru* I	GAATN$_7$T<u>GATC</u>GCAG	11.7	11,000,000
M.*Eco*D XI/M.*Taq* I	GAATN$_7$T<u>GATC</u>GA	7.9	57,000
M.*Eco*D XI/M.*Taq* II	GAATN$_7$T<u>GATC</u>GGTC	11.7	11,000,000
(M.*Eco*D XI/M.*Xba* I)	GAATN$_7$T<u>GATC</u>TAGA	11.7	11,000,000

Table 3 (con't)

Methylases	Recognition Sequence[#]	ERSS[@]	AFS
*M.Cla I	ATCGATCGAT	10.0	1,000,000
*M.Cla I/M.Mbo II	ATCGATCTTC	9.0	260,000
M.Cla I/M.Mme I	ATCGATCCRAC	9.4	470,000
M.Cla I/M.Nru I	ATCGATCGCGA	9.7	700,000
M.Cla I/M.Taq I	ATCGATCGGTC	9.7	700,000
(M.Cla I/M.Xba I)	ATCGATCTAGA	9.7	700,000
*M.Mbo II	GAAGATCTTC	10.0	1,000,000
M.Mbo II/M.Mme I	GAAGATCCRAC	9.4	470,000
M.Mbo II/M.Nru I	GAAGATCGCGA	9.7	700,000
*M.Mbo II/M.Taq I	GAAGATCGA	7.7	43,000
M.Mbo II/M.Taq II	GAAGATCGGTC	9.7	700,000
(M.Mbo II/M.Xba I)	GAAGATCTAGA	9.7	700,000
M.Mme I	GTYGGATCCRAC	11.0	4,200,000
M.Mme I/M.Nru I	GTYGGATCGCGA	10.4	4,900,000
M.Mme I/M.Taq I	GTYGGATCGA	7.0	57,000
M.Mme I/M.Taq II	GTYGGATCGGTC	10.4	1,900,000
M.Mme I/M.Xba I	GTYGATCTAGA	10.4	1,900,000
M.Nru I	TCGCGATCGCGA	12.0	16,800,000
M.Nru I/M.Taq I	TCGCGATCGA	7.9	57,000
M.Nru I/M.Taq II	TCGCGATCGGTC	9.7	700,000
(M.Nru I/M.Xba I)	TCGCGATCTAGA	11.0	4,200,000
*M.Taq II	TCGATCGA	8.0	66,000
M.Taq I/M.Taq II	TCGATCGGTC	7.9	57,000
(M.Taq I/M.Xba I)	TCGATCTAGA	7.9	57,000
M.Taq II	GACCGATCGGTC	12.0	16,800,000
(M.Taq II/M.Xba I)	GACCGATCTAGA	11.0	4,200,000
(M.Xba I)	TCTAGATCTAGA	12.0	16,800,000

*These cutting systems have been demonstrated [17,18]. The others are theoretical, based on known restriction recognition sequences. M.Hph I (GGTGA) has been excluded because it is now known to be a cytosine methylase (Feehery R. and Nelson M., unpublished results). Schemes using M.EcoB have not been tried but are likely to work as M.EcoB has the correct specificity (TG^{m}AN$_8$TGCT) [19]. However, both EcoB and EcoD XI are Type I modification methylases which make their use in vitro impractical. Acc III, Nru I and Xba I do not cleave DNA methylated at adenine in their recognition sequences [11,12,12a], a prerequisite if the corresponding methylase is to have adenine specificity. However, DNA from the species that carries Xba I is not detectably methylated at Adenine (M. Erlich, unpublished). The Acc III isoschizomer BspM II is not sensitive to adenine methylation and thus excluded (R. Morgan and I. Schildkraut, unpublished results). Isoschizomers of Dpn I include Cfu I, NmuE I, NmuD I and NsuD I (Camp R., Hurlin P. and Schildkraut I., unpublished results). Isoschizomers of Mbo II include Nsu I. Isoschizomers of Nru I include Ama I and Sbo 13 [1].

Table 3 (con't)

#Where there is more than one recognition sequence for *Dpn* I only one of the combinations is shown.

@The effective size is the log (base4) of the sum of the cutting frequency per base pair. For example, for M.*Cla* I/M.*Mbo* II this is calculated from the frequency of GAAGATCTTC ($1/4^{10}$) + GAAGATCGAT ($1/4^{10}$) + ATCGATCTTC ($1/4^{10}$) = ($1/4^9$). Effective size = 9 base pairs.

The difference in the sensitivity of these enzymes to DNA methylation can be utilized to enhance the apparent specificity of the endonuclease:

(1) M.*Hpa* II (C^mCGG) methylation.

    Result: GGATC^{m}C occurs at the overlapping sequence

        5' N C^{m}C G G A T C C N 3'   [A]
        3' N G G^{m}C C T A G G N 5'

    and

        5' N C^{m}C G G A T C^{m}C G G N 3'   [B]
        3' N G G^{m}C C T A G G^{m}C C N 5'

(2) M.*Bam*H I methylation.

    Result: All GGATCC sequences are methylated at GGAT^{m}CC except those that overlap GGCC (sequences [A] and [B]). Since M.*Bam*H I is a monomer that interacts with only half of the double-stranded DNA, the sequence [A] is methylated by M.*Bam*H I in one strand (hemimehtylation) to produce

        5' N C^{m}C G G A T^{m}C C N 3'
        3' N G G^{m}C C T A G G N 5'

    However, the sequence [B] is protected from M.*Bam*H I in both strands.

(3) *Bam*H I cleavage.

    Result: The sequence [B], C^{m}CGGATC^{m}CGG, remains available for *Bam*H I cleavage. Hemimethylation of one strand at GGAT^{m}CC in sequence [A] is sufficient to block cleavage. Thus, all *Bam*H I recognition sequences except [B] are protected from cleavage by methylation in at least one strand by M.*Bam*H I.

    Conclusion: The *Bam*H I cleavage specificity is enhanced to 10 base pairs (CCGGATCCGG).

Examples of methylase/endonuclease pairs we have tested that have the requisite properties to be rare cleavage systems are M.*Hpa* II (C^mCGG)/M.*Bam*H I/*Bam*H I which gives cleavage at CC(A/T)GGATCC(A/T)GG and M.*Ava* I (^mCYCGRG)/M.*Bam*H I/*Bam*H I which gives cleavage at CYCGRGGATCCYCGRG (Table 4).

TABLE 4. Methylase/M.*Bam*H I/*Bam*H I Recognition Sequences

METHYLASE(S)	METHYLASE SPECIFICITY	NEW CLEAVAGE SPECIFICITY	ERSS
M.*Hpa* II	CmCGG	CCGGATCCGG	10.0
M.*Msp* I	mCCGG	CCGGGATCCCGG	12.0
M.*Dde* I	mCTNAG	CTNAGGATCCTNAG	12.0
M.*Ava* I (M.*Aqu* I)	mC(^C_T)CG(^A_G)G	mC(^T_C)CG(^A_G)GGATCC(^C_T)CG(^G_A)G	14.0
Mammilian mCG Methylase	mCG	CGGATCCG	8.0
M.*Hpa* II + M.*Msp* I	CmCGG / mCCGG	CCGGATCCCGG	9.7
M.*Hpa* II + M.*Dde* I	CmCGG / mCTNAG	CCGGATCCTNAG	9.9
M.*Hpa* II + M.*Ava* I	CmCGG / mC(^C_T)CG(^G_A)G	CCGGATCC(^C_T)CG(^G_A)G	9.9
M.*Msp* I + M.*Dde* I	mCCGG / mCTNAG	CCGGGATCTNAG	11.0
M.*Msp* I + M.*Ava* I	mCCGG / mC(^C_T)CG(^G_A)G	CCGGGATCC(^C_T)CG(^G_A)G	11.7
M.*Dde* I + M.*Ava* I	mCTNAG / mC(^C_T)CG(^G_A)G	CTNAGGATCC(^C_T)CG(^G_A)G	11.9
M.*Dde* I + M.*Ava* I	mCTNAG / mC(^C_T)CG(^G_A)G	CTNAGGATCC(^C_T)CG(^G_A)G	11.9
M.*dcm* (M.*Eco*R II)	CmC(^T_T)GG	CC(^A_T)GGATCC(^A_T)GG	11.0
M.*dcm* + M.*Dde* I	CmC(^A_T)GG / mCTNAG	CC(^A_T)GGATCCTNAG	10.7
M.*dcm* + M.*Hpa* II	CmC(^A_T)GG / CmCGG	CC(^A_T)GGATCCGG	9.7

TABLE 4. (Continued)

METHYLASE(S)	METHYLASE SPECIFICITY	NEW CLEAVAGE SPECIFICITY	ERSS
M.*dcm* + M.*Msp* I	CmC(^A_T)GG mCCGG	CC(^A_T)GGATCCCGG	10.7
M.*Ava* I + M.*dcm*	mC(^C_T)CG(^G_A)G CmC(^A_T)GG	C(^C_T)GG(^G_A)GGATCC(^A_T)GG	9.7

In general, if (i) a specific DNA methylation completely blocks cutting by an endonuclease and (ii) the methylase is sensitive to prior methylation at a non-cognate site but the endonuclease is not, then a rare cleavage system can be devised along the lines we have outlined. We predict that there are about 500 different cleavage specificities possible in the 8 to 18 base pair range that can be devised for the 400 known restriction systems [1].

V. NON-RANDOM SEQUENCES IN NATURAL DNA

For <u>random</u> DNA a restriction endonuclease with a six base recognition sequence will cut on average once every 4,096 base pairs. In this section we will take advantage of the <u>non-random</u> arrangement of natural DNA sequences to predict which restriction endonucleases will cut more rarely to produce larger fragments. Such non-random character in DNA can be uncovered from mononucleotide, dinucleotide and trinucleotide frequencies.

a. Mononucleotides

G+C contents have been calculated for the genomes of a large number of species [6,20]. These vary from 22% G+C to 73% G+C (Table 5). Most thermophiles have a G+C content in excess of 60%. When studying a DNA which is known to be very asymmetric in base composition, one can select restriction endonucleases which have recognition sequences that include the rare bases as good candidates for rare cleavage. Table 5 shows the effect of base composition on the AFS for a number of restriction endonucleases. For example, *Sma* I (CCCGGG) should cut a genome of 30% G+C, such as plasmodium or mycoplasma, once every 88,000 base pairs (ERSS 8.2 base pairs).

TABLE 5. % G+C in the genomes of various species.
From [6].   M = mitochondria; C = chloroplast

Species	<40%	% G+C 40-60%	>60%
Bacteria			
Thermus aquaticus			66
Bacillus subtilis		45	
Rhodopseudomonas spheroides			69
Mycoplasma sp	25-35		
Escherichia coli		50	
Actinomyces sp			60-73
Clostridium sp	25-40		
Vibrio cholera		46	
Streptomyces sp			70-77
Micrococcus sp			60-64
Mycobacterium tuberculosis			64
Rickettsia rickettsi	32		
Rhizobium sp			57-63
Agrobacterium sp			58-66
Caulobacter sp			60-67
Arthrobacter sp			60-72
Streptococcus sp	34-43		
Myxococcus sp			67-71
Nocardia sp			62-73
Pseudomonas aeruginosa			66
Algae			
Chlamydomonas reinhardi	(38C)		65
Nostoc punctiform		44	
Spyrogyra sp		40	
Fungi			
Aspergillus sp		52-58	
Candida albicans	35		
Dictyostelium discoidaeum	22		
Neurospora crassa		54	
Penicillium sp		49-54	
Saccharomyces cerevisiae	(18M)	40	
Schizosaccharomyces pombe		41	
Protozoa			
Euglena graclis	(25M) (26C)	50-53	
Leishmania sp		54-57	
Plasmodium		40	
Tetrahymena sp	19-32		
Trypanosome sp		45-50	
Amoeba proteus			66
Porifera		36-60	
Coelenterates	37-42		
Mollusca	32-47		
Annelidia		40	
Arthropods			
Drosophila melanogaster		40	
Apis mellifica	33		

Table 5 (continued)

Species	<40%	% G+C 40-60%	>60%
Echinodermata	35-46		
Vertebrates		40-44 (41M)	
Gymnosperms			
Ginko	35		
Pinus		40	
Angiosperms			
Zea mays		41 (39C)	
Pea	37 (38C)		
Cotton	32 (37C)		

TABLE 6. Dependance of restriction endonuclease cleavage frequency on base composition.

The average fragment size (AFS) predicted for a restriction recognition sequence depends on base composition. In this table a range of 30% to 70% G+C content and a random distribution of bases were assumed. The predicted cleavage frequencies were then calculated for two types of restriction and recognition sequences, one containing six cytosines plus guanines and one containing two adenines plus thymines and four cytosines plus guanines. ERSS = Effective recognition sequence size.

	Sma I (CCCGGG)		BamH I (GGATCC)	
% G+C	AFS	ERSS	AFS	ERSS
70	500	4.5	3,000	5.8
60	1,400	5.2	3,100	5.8
50	4,100	6.0	4,100	6.0
40	16,000	7.0	7,000	6.4
30	88,000	8.2	16,000	7.0

TABLE 7. Effect of dinucleotide usage in *Saccharomyces cerevisiae* DNA from GENBANK were analyzed for the observed number of restriction endonuclease cleavage sites. This number was compared to the cleavage frequency expected from dinucleotide usage [6,20]. The rarest dinucleotides are GC (1.7%), CG (1.8%), GG (2.3%), and CC (2.3%).
AFS = Average fragment size; ERSS = Effective recognition sequence size.

Restriction Sequence	Enzyme Name	Observed No. of cuts	Expected	ERSS	AFS
GCGCGC	*Bss*H I	4	5.2	7.5	33,000
CCGCGG	*Sac* II	5	5.6	7.3	27,000
CGGCCG	*Eag* I	9	5.6	6.9	15,000
GGCGCC	*Nar* I	10	7.1	6.8	13,000
GCCGGC	*Nae* I	4	7.1	7.5	33,000
CCCGGG	*Sma* I	4	7.8	7.5	33,000
GGGCCC	*Apa* I	11	9.6	6.8	12,000
CCTAGG	*Avr* II	2	16.3	8.0	67,000
GGATCC	*Bam*H I	9	20.0	6.9	15,000
GAATTC	*Eco*R I	46	54.7	5.7	3,000
TTTAAA	*Dra* I	78	136.0	5.3	2,000

b. Dinucleotides

Nearest neighbor frequencies have been calculated for the DNA of a large number of species [6,20]. At this second level of complexity (Table 7) one can select restriction endonucleases which have recognition sequences containing one or more rare dinucleotides. Table 6 gives an example of this strategy using the *S. cerevisae* genome. The restriction endonuclease *Sma* I should cut this DNA once every 33,000 base pairs (ERSS 7.5 base pairs).

c. Trinucleotides

The sequences of some portions of DNA from over 100 species are now available (GENBANK). Statistically significant trinucleotide frequencies can be obtained when about 20,000 base pairs of representative DNA from a species, genus or family are available. Trinucleotides often show dramatic differences in their frequencies, particularly in organisms with small genome sizes such as prokaryotes and algae, because of the effects of codon usage.

Table 8 gives an example of rare cleavage sites predicted from rare trinucleotides in 20,000 base pairs of *E. coli* DNA. The frequency of each of the predicted rare cleavage sites was then determined for 290,000 base pairs of *E. coli* DNA sequence (GENBANK). It can be seen that many of the predicted rare cutters should cut with a much lower frequency than once every 4^6 base pairs. In some cases the predicted AFS is 27,000 base pairs, equivalent to a 7.4 base pair cleavage specificity.

Figure 1 shows a Pulsed Field Gel of *E. coli* DNA digested with *Spe* I, *Nhe* I, *Mst* II, *Eco*R I, and *Hin*d III. It can be seen that the number of fragments below the lambda markers (first and last lanes) are substantially less for the three TAG containing restriction recognition sequences than for *Eco*R I and *Hin*d III, indicating that *Spe* I, *Nhe* I and *Mst* II recognition sequences are two-to-ten fold rarer than sequences that do not contain TAG.

Predicted rare cutters for various species are given in Table 9. It should be noted that trinucleotides containing CpG are rare in a number of species, e.g. vertebrates and higher plants. The rarity of CpG is discussed below. Furthermore, TAG and its complement, CTA, are rare in DNA that has been sequenced from many species. This effect is enhanced in species in which most of the DNA codes for proteins. Part of the reason why these trinucleotides are rare in coding regions is that TAG is a stop codon and must, therefore, be discriminated against in the open reading frame, |TAG|, but not necessarily in other frames; |nnT|AGn| and |nTA|Gnn|. In the case of CTA, |nnC|TAa| and |nnC|TAg| would be stop codons too, but |CTA|, |nnC|TAy| and |nCT|Ann| might not be discriminated against. However, we have determined that TAG and CTA are at least two-fold rarer than expected in all reading frames in many species. Why this is the case is unknown, but it does explain the extreme rarity of TAG/CTA in species with a high proportion of protein coding sequences.

A number of restriction endonucleases contain more than one trinucleotide that is rare in *E. coli* DNA within their recognition sequences, e.g. *Avr* II (CCTAGG), *Spe* I (ACTAGT) and *Xba* I (TCTAGA). We expect these restriction endonucleases to be especially useful in pulsed-field electrophoretic mapping of DNAs from many prokaryotes.

d. Best Predictor

In many cases entire phyla display the same distributions in dinucleotide frequency (21,22) and trinucleotide frequency (23). Therefore it is often not necessary to have sequence information for a particular species in order to predict rare restriction targets.

TABLE 8. Effect of trinucleotide usage in *E. coli* on the predicted cleavage frequency of selected restriction endonucleases.

The average frequency of each trinucleotide in a sample of 313,424 base pairs of *E. coli* DNA was 4897. The rarest trinucleotides were TAG (1707), CTA (2316), CCC (2972), CTC (3203), GAG (3231), AGT (3327), GGG (3475), CCT (3500) and TAC (3579). On average, restriction recognition sequences of six base pairs occurred 76 times in this sample.

Note that *Avr* II and *Spe* I recognition sequences are particularly rare because they contain both TAG and one other rare trinucleotide. Percent of expected is observed divided by 76. AFS = Average fragment size calculated by division of 313,424 by the observed number of recognition sequences; ERSS = effective recognition sequence size calculated as the log (base 4) of the AFS.

Rare Tri-nucleotide	Restriction Site	Name	Frequency	% of Expected	AFS	ERSS
TAG/	TAGCTA	none	25	33	12,500	6.8
CTA	CTATAG	none	18	24	17,400	7.0
	ACTAGT	*Spe* I	15	20	21,000	7.2
	CCTAGG	*Avr* II	8	10	39,000	7.6
	GCTAGC	*Nhe* I	11	15	28,500	7.4
CCC/	CCCGGG	*Sma* I	46	60	6,800	6.4
GGG	GGGCCC	*Apa* I	8	10	39,000	7.6
CTC/	CTCGAG	*Xho* I	24	31	13,100	6.8
GAG	GAGCTC	*Sac* I	18	24	17,400	7.0
AGT/	AGTACT	*Sca* I	32	42	9,800	6.5
ACT	ACTAGT	*Spe* I	15	20	21,000	7.2
CCT/	CCTAGG	*Avr* II	8	10	39,000	7.6
AGG	AGGCCT	*Stu* I	45	59	7,000	6.4
	CCTNAGG	*Mst* II	23	30	13,600	6.9
TAC/	TACGTA	*Sna*B I	50	65	6,300	6.3
GTC	GTCTAG	*Sal* I	36	47	8,700	6.5
none	GGATCC	*Bam*H I	47	61	6,700	6.4
	AAGCTT	*Hin*d III	55	72	5,700	6.2
	GAATTC	*Eco*R I	66	86	4,700	6.1
	TTTAAA	*Dra* I	121	158	2,600	5.7

FIG. 1. Pulsed Field Gel electrophoresis, 1% agarose, 200V, 2 second pulse, 12 hours.

Lane 1: Lambda DNA marker (48,000 base pairs)

Lane 2: *Spe* I digest of *E. coli* DNA

Lane 3: *Nhe* I digest of *E. coli* DNA

Lane 4: *Mst* II digest of *E. coli* DNA

Lane 5: *Hin*d III digest of *E. coli* DNA

Lane 6: *Eco*R I digest of *E. coli* DNA

Lane 7: Lambda DNA marker

TABLE 9. Rarest and most common trinucleotides in sequenced DNA from a variety of species.

For particular species the rare restriction endonuclease recognition sequences include the trinucleotides listed in column 1 to 5. These frequencies were calculated from DNA sequences in GENBANK. The trinucleotide frequency in % is shown below each pair, in some cases. The frequency of a typical trinucleotide pair in DNA is 1/32 = 3.13%. Where possible, for each species the trinucleotide frequencies were made more representative of their natural abundance by altering the ratio of protein coding, non-coding and repetitive DNA in the sample to reflect the actual abundance of these types of DNA in the genome. Also, note that TAG/CTA is rare in protein coding regions, thus *Avr* II (CCTAGG) and *Spe* I (ACTAGT) are likely to be rare cutters in *E. coli* chromosomal DNA and other genomes that have a high proportion of protein coding DNA. ACG/CGT and TCG/CGA are rare in vertebrates and plants, thus, restriction sites that contain these trinucleotides are also rare (Table 9).

SPECIES	RAREST 1	2	3	4	5	COMMONEST
Mammals	CGT/ACG 0.94	TCG/CGA 1.00	CGC/GCG 1.36	CCG/CGG 1.54	GTA/TAC 1.87	TTT/AAA CTG/CAG
Birds	TCG/CGA	CGT/ACG	CTA/TAG	CGC/GCG	GTA/TAC	TTT/AAA CTG/CAG
Amphibia	CGT/ACG	CTA/TAG	GTA/TAC	TCG/CGA	ATA/TAT	(CCC/GGG GCC/CGG)
Fish	CTA/TAG	TCG/CGA	CGT/ACG	GTA/TAC	GCG/CGC	TTT/AAA CTG/CAG
Sea Urchin	TAG/CTA	GCG/CGC	CGG/CCG	CGT/ACG		TTT/AAA ATT/AAT
Drosophila	GGG/CCC 1.96	CGG/CCG 2.03	CGT/ACG 2.04	GCG/CGC 2.19	TAG/CTA 2.19	TTT/AAA ATT/AAT
B. mori	GGG/CCC	CCT/AGG	TAG/CTA	CGT/ACG		CAG/CTG
Aplysia	GCG/CGC	CCG/CGG	GGC/GCC	TAG/CTA		AAA/TTT
Trypanosomes	ATC/GAT	GTA/TAC	GGG/CCC	CGT/ACG		TTT/AAA
Tetrahymena	CGT/ACG	CGC/GCG	CCC/GGG	GCC/GGC		TTT/AAA
Dictostyleum	CTA/TAG	GCA/TGC	CGC/GCG	CAG/CTG		TTT/AAA
S.cerevisiae	CGC/GCG 1.20	CCG/CGG 1.40	CCC/GGG 1.38	GGC/GCC 1.65	CGT/ACG 1.85	TTT/AAA ATT/AAT
N. crassa	TAG/CTA	ATA/TAT	TTA/TAA	GTA/TAC		TTT/AAA AAT/ATT

Table 9 (continued)

SPECIES	RAREST 1	2	3	4	5	COMMONEST
Flowering Plants	GCG/CGC	CGG/CCG	ACG/TCG	CGA/TCG	GGG/CCC	TTT/AAA CAA/TTG
Bacillus	CTA/TAG	CCC/GGG	ACT/AGT	CGT/ACG	CAC/GTG	AAA/TTT GAA/TTC
E. coli	TAG/CTA 1.28	CTC/GAG 2.05	CCC/GGG 2.06 19	CCT/AGG 2.28	AGT/ACT 2.29	AAA/TTT CTG/CAG

One might imagine that the best predictor of the cleavage frequency of a particular enzyme is the frequency of its recognition sequence in a DNA sequence sample. However, unless the sample is large (>20,000 base pairs), then the better predictor is trinucleotide frequency. For instance, in a randomly chosen sample of 20,000 base pairs of *E. coli* DNA, the sequence AGTACT (*Sca* I) occurred 3 times despite the fact that *Sca* I is a predicted rare cutter. In contrast, the sequence GAATTC (*Eco*R I) occurred zero times despite the fact that *Eco*R I is a predicted frequent cutter. Calculations using trinucleotides from the same sample give a much more accurate estimate of the overall frequency of these hexanucleotides in the genome.

Methods for calculating the expected frequency of a six base sequence is given in references 22 and 24. In brief, the frequency of the hexanucleotide sequence 5' UVWXYZ 3' is:

$$\frac{P(UVW) \cdot p(VWX) \cdot p(WXY) \cdot p(XYZ)}{p(UW) \cdot p(WX) \cdot p(XY)}$$

where U thru Z are each one of the four bases A, C, G, or T and where p(UVW) is the probability of the sequence, 5' UVW 3', in the sample. This algorithm is essentially a function of both di- and trinucleotide frequencies in all reading frames.

For example, the calculated AFS of CCTAGG (*Avr* II) in mammalian DNA is:

$$\frac{p(CCT) \cdot p(CTA) \cdot p(TAG) \cdot p(AGG)}{p(CT) \cdot p(TA) \cdot p(AG)} = \frac{0.0111 \times 0.00073 \times 0.0054 \times 0.0117}{0.0553 \times 0.0457 \times 0.0533}$$

=1/25,700. In other words the predicted AFS is 25,700 base pairs.

From a practical stand point one may choose from a large number of cleavage specificities whose few cutting strategies which are likely to produce average fragment sizes in the range of interest.

e. Intragenomic Sequence Variation

Considerable intragenomic variation in DNA sequences can occur. A knowledge of this level of asymmetry can be applied in some cases to give a more accurate picture of which recognition sequences are rare. For example, in mammalian DNA the rarity of CpG (1% versus 4% expected) and the A+T richness (60%) of the genome were used together to predict the AFS of CpG-containing restriction recognition sequences in mammalian DNA (Table 10). However, when the frequencies of these restriction endonuclease recognition sequences were determined for 918,790 base pairs of sequenced mammalian DNA from GENBANK, some hexanucleotides that contain A+T and CpG, such as TCGCGA (Nru I), were rarer than hexanucleotides containing G+C and CpG such as CCCGGG (Sma I), or even the eight base recognition sequences of Sfi I GGCCN$_5$GGCC and Not I GCGGCCGC (which has two CpGs). This expected result may be explained by a further level of asymmetry in the mammalian genome: CpG sequences tend to occur in G+C rich regions ([21,25]. Therefore, CpG sequences adjacent to A or T might be exceedingly rare in mammalian DNA. However, this data should be treated with caution since the sample of 918,790 base pairs of sequenced DNA is very atypical of the total genome.

f. Other considerations when Predicting Rare Cutters from Sequence Information

Caution should be used when applying the strategies we have described to identify rare restriction sites. For example, the rarity of CpG in mammalian DNA is probably due to the hypermutability of ^{m}GpG. Since many restriction endonucleases are sensitive to methylation at cytosine and perhaps 90% of CpG sites are methylated or partially methylated in mammalian DNA, it follows that many enzymes with recognition sequences containing CpG will not cut mammalian DNA to completion. Nevertheless, a large number of CpG rich, G+C rich islands have been observed in coding regions of mammalian DNA and these have been methylation free [25]. Thus, restriction endonucleases which have recognition sequences that contain CpG may be useful for dissecting these regions out of the genome.

There are likely to be other cases where di- or trinucleotide sequences are rare because they are the targets of a sequence-specific cytosine methylase and thus hypermutable. In such cases some of the restriction endonucleases selected by these methods are likely to be methylation-sensitive and may not cleave the DNA at all. The sensitivity of several restriction endonucleases to site-specific methylation has been compiled [9,26]. Furthermore, some enzymes cut very poorly at certain cleavage sites. These enzymes may not cut DNA to completion under any circumstances. These include Nar I, Nae I and Sac II.

TABLE 10. Effect of CpG frequency in mammals on the predicted cleavage frequency of selected recognition sequences.

The dinucleotide and trinucleotide frequencies were determined for a sample of 918,790 base pairs of mammalian DNA which excluded alleles and mRNAs. The expected frequency for the recognition sequences of commercially available restriction endonucleases and other cleavage systems were calculated [21,22,23]. The observed frequencies of these sites in the sample are underlined. Tri = Predicted number based on trinucleotide frequency [17]. AFS = Average fragment size expected from the total dinucleotide frequencies in the genome [6].

* No CpGs *

A+T Content	Enzyme	Recognition Sequence	Obs	Tri	AFS
6	Dra I	TTTAAA	582	134	1,500
6	Ssp I	AATATT	408		2,400
4	EcoR I	GAATTC	250	81	2,800
4	HinD III	AAGCTT	225	71	3,800
2	Avr II	CCTAGG	232	220	4,500
2	BamH I	GGATCC	211	108	4,800
0	Apa I	GGGCCC	317	86	6,600

* 1 CpG *

4	Asu II	TTCGAA	48	64	9,900
4	Cla I	ATCGAT	40	114	15,000
4	SnaB I	TACGTA	32	131	31,000
3	Fsp I	TGCGCA	87	60	17,000
3	Xho I	CTCGAG	175	116	17,000
3	Aat II	GACGTC	71	170	30,000
3	Sal I	GTCGAC	51		30,000
0	Sma I	CCCGGG	344	176	26,000
0	Nae I	GCCGGC	175	154	36,000
0	Nar I	GGCGCC	202	132	36,000

* 2 CpGs *

2	Nru I	TCGCGA	26	17	93,000
2	Pvu I	CGATCG	28	162	103,000
2	Mlu I	ACGCGT	21	72	130,000
0	Sac II	CCGCGG	199	269	140,000
0	Eag I	CGGCCG	185	406	140,000
0	BssH I	GCGCGC	180	210	200,000

* Other rare cutters *

	Enzyme	Recognition Sequence	Obs		AFS
	Sfi I	GGCCN$_5$GGCC	34		150,000
	Rsr II	CGG(A/T)CCG	23		260,000
M.Mbo II/Dpn I		GAAGATCTTC	1		530,000
M.Taq I/Dpn I		TCGATCGA	1		1,100,000
	Not I	GCGGCCGC	40		3,300,000
M.Cla I/Dpn I		ATCGATCGAT	0		17,000,000

VI. GENE MAPPING AT THE MEGABASE LEVEL

The non-random nature of genomic DNAs leads to species-dependent alterations in the frequencies of restriction and methylase/endonuclease recognition sequence combinations. In the methylase cross-protections shown in Table 2 and the methylase/Dpn I systems shown in Table 3 the AFS and ERSS are calculated for totally random DNA. For example, the dam methylation enhanced Cla I recognition sequence (A/T/C)ATCGAT(A/T/G) has a calculated AFS of 7,512 base pairs on random DNA, but an AFS of 26,000 base pairs is predicted by dinucleotide frequency in mammalian DNA. The M.Nru I/Dpn I recognition sequence TCGCGATCGCGA has a calculated AFS of 1.7×10^7 base pairs on random DNA, but because this sequence contains two CpGs it may not occur at all in the 3×10^9 base pairs of the human genome (see Table 11).

Many hexamer restriction endonucleases and methylase/endonuclease combination cleavage schemes are thus suitable for physical mapping of chromosomes from the 100 kilobase to the megabase level, depending upon the DNA source.

In the past, mapping on this scale has been measured genetically in recombination units (centiMorgans). One centiMorgan is defined as a one per cent recombination frequency per generation between two genetically linked markers. Genetic recombination frequencies vary between different organisms. In E. coli one centiMorgan (cM) equals about 1,500 base pairs; in S. cerevisiae one cM equals about 2,500 base pairs; in humans one cM is approximately 900,000 base pairs. Therefore physical lengths (measured in kilobases or megabases) and genetic distances (measured in centiMorgans) must be calibrated separately for different species. Table 11 displays graphically the relative frequencies of several different DNA cutting schemes for Drosophila melanogaster and human genomes. From this diagram it can be seen that different enzymatic techniques are more suitable for chromosome mapping in each organism. If the genetic distance between two markers is known, then the approximate physical distance between them can be calculated: appropriate DNA cutting schemes may be chosen so that physical linkage maps may be constructed.

VII. SUMMARY

We have described how DNA methylases may be used to enhance the apparent specificities of restriction endonucleases [9-11,17,18] to generate DNA fragments averaging 6000 to 270,000,000 base pairs on random DNA. Taking into account the non-random arrangement of natural DNA, we have predicted the rarity of certain recognition sequences in the genomes of several species.

TABLE 11. Gene mapping at the megabase level.

The columns are: A log (base 4) scale of DNA size in base pairs; A log (base 10) scale of DNA size in base pairs; Average fragment size (AFS) produced in Drosophila DNA by various potential cleavage systems; The recombination distance in centiMorgans for Drosophila; Average fragment size (AFS) produced in Human DNA by various potential cleavage systems; The recombination distance in centiMorgans for Humans.

Note that the exponent in the log (base 4) scale in the effective recognition sequence size. A centiMorgan is the genetic distance that gives a 1% recombination rate between markers per generation. For Drosophila one centiMorgan is approximately 250,000 base pairs. For Humans it is approximately 900,000 base pairs.

Base Pairs		Drosophila ERSS	cM	Humans ERSS	cM
4^{16}	10^{10}			M.*Nru* I/*Dpn* I	
				<<<Human Genome Size	
4^{15}	10^{9}				400
4^{14}					
		<<<Drosophila Genome Size			
				M.*Cla* I/*Dpn* I	
4^{13}	10^{8}		400	M.*Acc* III/*Dpn* I	40
4^{12}		M.*Bsp*H I/*Dpn* I		M.*Bsp*H I/*Dpn* I	
		M.*Nru* I/*Dpn* I			
	10^{7}		40		4
4^{11}					
		M.*Cla* I/*Dpn* I			
				Not I	
4^{10}	10^{6}	M.*Mbo* II/*Dpn* I	4		1
				M.*Mbo* II/*Dpn* I	
				Rsr II	
4^{9}		*Not* I		*Bss*H II	
	10^{5}	*Sfi* I	1	*Sfi* I	
		Rsr II		*Pvu* I	
4^{8}				*Nru* I	
4^{7}		*Apa* I, *Sma* I		*Fsp* I, *Xho* I	
	10^{4}	*Fsp* I, *Avr* II		*Cla* I, *Apa* I	
		Cla I			

Together, these methods allow for rational selection of tools for the enzymatic dissection of DNA molecules ranging in size from large viruses (100,000 base pairs) to large chromosomes (400,000,000 base pairs).

VIII. ACKNOWLEDGEMENTS

We would like to thank David Glass and Mike Lane for critical reading of the manuscript. Michael McClelland is a Lucille P. Markey Biomedical Scholar and this work was supported by the Lucille P. Markey Charitable Trust. Michael Nelson is a Ruth M. Reeves Fellow.

IX. REFERENCES
1. Roberts, R.J. (1985) Nucleic Acids Res. 13: r165-r200.
2. Schwartz, D. and Cantor C.R. (1984) Cell 37: 67-75.
3. Carle, G.F. and Olsen M.V. (1984) Nucleic Acids Res. 12: 5647-5664.
4. Carle, G.F., Frank, M., and Olsen, M.V. (1986) Science 232: 65-69.
5. McClelland, M. (1986) Methods in Enzymology: Recombinant DNA (in press).
6. Normore, W.M., Shapiro, H.S., and Setlow, P. (1976) CRC Handbook of Biochemistry and Molecular Biology, (Ed. G.D. Fasman), CRC Press.
7. Smith, H.O. (1979) Science 205: 455-462.
8. Janulaitis, A., Kimasauskas, S., Petrusyte, M., and Butkus, V. (1983) FEBS Lett. 161: 131-134.
9. McClelland, M. and Nelson, M. (1985) Nucleic Acids Res. 13: r201-207.
10. McClelland, M. (1981) Nucleic Acids Res. 9: 6795-6804.
11. Nelson, M., Christ, C., and Schildkraut, I. (1984) Nucleic Acids Res. 12: 5165-5173.
12. Nelson, M. and Schildkraut, I. (1986) Methods in Enzymology, Recombinant DNA (in press).
12a. Kita, K., Nobutsuga, H., Aksushi, O., Kadonoshi, S., and Obayashi, A. (1985) Nucleic Acids Res. 13: 8685-8693.
13. Kaui, L. and Piekarowicz, A. (1978) Eur. J. Biochem. 92: 417-426.
14. Schlagman, S.L. and Hattman, S. (1983) Gene 22: 139-156.
15. Lacks, S. and Greenberg, B. (1980) J. Mol. Biol. 114: 153-168.
16. Geier, G.E. and Modrich, P. (1979) J. Biol. Chem. 254: 1408-1413.
17. McClelland, M., Kessler, L. and Bittner, M. (1984) Proc. Natl. Acad. Sci. USA 81: 983-987.
18. McClelland, M., Nelson, M. and Cantor, C. (1985) Nucleic Acids Res. 13: 7171-7182.
19. van Ormondt, H., Lautenberger, J.A., Linn, S. and deWaard, A. (1973) FEBs Lett. 33: 177-180.
20. Nussinov, R. (1984) Nucleic Acids. Res. 12: 1749-1762.
21. McClelland, M. and Ivarie, R. (1982) Nucleic Acids Res. 10: 7865-7877.
22. McClelland, M. (1985) J. Mol. Evol. 21: 317-322.
23. Grantham, R., Gautier, C., Gouy, M., Jacobzone, J. and Mercier, R. (1981) Nucleic Acids Res. 9: r43-r74.

24. Waterman, M.S. (1983) Nucleic Acids Res. 11: 8951-8956.
25. Bird, A., Taggart, M., Frommer, M., Miller, O.J., and Macleod, D. (1985) 40: 91-100.
26. Kessler, C., Neumaier, P.S., and Wolf, W. (1985) Gene 33: 1-102.
27. Roy, P.H. and Smith, H.O. (1973) J. Mol. Biol. 81: 445-459.
28. Hattman, S., Keister, T. and Gotteher, A. (1978) J. Mol. Biol. 125: 701-711.
29. Nur, I., Szyf, M., Razin, A., Glaser, G., Rottman, S. and Razin, S. (1985) J. Bact. 164: 19-24.
30. Wagner, H., Simon, D., Werner, E., Gelderblom, H., Darai, C. and Flugel, R. (1985) J. Virol. 53: 1005-1007.
31. Willis, D., Goorha, R. and Granoff, A. (1984) J. Virol. 49: 86-91.
32. Braumberg, S., Pratt, K. and Hattman, S. (1982) J. Bact. 150: 993-996.

Index

AacI, 3

AaeI, 3

AamI, 6

Aat endonucleases, 3

AbrI, 6

AcaI, 5

Acc endonucleases, 4

ACCI promoter, 221-224

AcyI, 5

S-Adenosyl-L-methionine, 148

Afl endonucleases, 5

AFS. See Average fragment size

Aha endonucleases, 5-6

AimI, 3

Ali endonucleases, 3

AluI, 6

AmaI, 4

Amino acid
  residue requirements
    for BamH I
      endonuclease
      activity, 169-172
    for BamH I methylase
      activity, 173-174
  sequence, of BamH I
    endonuclease, 153

AniI, 6

Aoc endonucleases, 5

AorI, 3

Aos endonucleases, 5

ApaI, 3

ApaLI, 3

AprI, 4

ApuI, 5

ApyI, 6

AquI, 4

Arginine, BamH I
  endonuclease activity
  and, 169-170

Asp endonucleases, 3-4, 4-5

AstWI, 5

Asu endonucleases, 5

Atu endonucleases, 4

Aua endonucleases, 5

Average fragment size
  (AFS)
  for BamH I, 270
  for methylases, 264-265
  production strategies, 258
  for SMA 1, 270
  of trinucleotides, 273

Aur endonucleases, 5

AxxaI, 3

BaaH II endonuclease, 251
BAAH II methylase, 251
Bac endonucleases, 6
Bacteriophage T4, 234
BalI, 11
BamH I.1 endonuclease, 175-179
BamH I deoxydinucleotide subsets
  inhibition constants with BamH I, 155, 156
  protection from butanedione modification and, 170-172
  synergistic inhibitory effects, 156, 157
BamH I endonuclease
  amino residues required for activity, 169-172
  average fragment size, 270
  catalytic properties, 155-158
  cleavage sites, 6
  cleavage specificity, 266
  dimethylsulfoxide and, 175-176
  DNA sequence, 6
  ERRSS, 270
  flanking sequences, 158
  glycerol and, 175-176
  hydrophobicity, 154
  inhibition with non-specific DNA, 168-169
  interaction with left-handed Z-DNA, 251-252
  kinetic mechanisms, 157-158
  methylation, 266
  microorganisms, 6
  perturbation of sequence specificity by organic solvents, 175-179
  physical relationship with BamH I.1, 175-179
  protection against butanedione

modification, 170–172
purification of, 148–150
reaction rates and relative location of BamH I sites, 163–164
recognition sequences, 266
site discrimination, 181
structural characteristics, 153–154
time course for cleavage of form I SV40 DNA, 158
BamH I methylase
amino acid residues, required, 173–174
catalytic properties, 158–162
inhibition with non-specific DNA, 168–169
perturbation of sequence specificity by organic solvents, 180–181
purfication of, 150–152
reaction rates and relative location of BamH I sites, 163–164
site discrimination, 181
specificity verification, 152
Vmax, 181
BamH I restriction-modification system, 148–181
Ban endonucleases, 6
BauI, 6
Bbe endonucleases, 10–11
Bbi endonucleases, 10
BbrI, 11
Bbu endonucleases, 6
Bce endonucleases, 7
BclI, 7
BcnI, 7
Bgl endonucleases, 7
Bin endonucleases, 11
BloI, 11
Blu endonucleases, 11

Bme endonucleases, 7
Bpel, 11
BprI, 11
Bpul endonucleases, 7
BsaPI, 9
BscI, 7
Bse endonucleases, 9
BsePI, 9
BsmI, 9
BsoPI, 9
Bsp endonucleases, 7-8
Bsr endonucleases, 9
Bss endonucleases, 8-9
Bst endonucleases, 8, 9
Bsu endonucleases, 9-10
Bth endonucleases, 11
BtiI, 10
Buffer systems, 87-89
Butanedione modification
  of Bam H I endonuclease
    and protective
    effect of
    deoxydinucleotide
    subsets, 170-172
  of BamH I methylase, 173-174
BuuI, 10

CaiI, 11

Catalytic properties, of
  BamH I endonuclease,
  155-158
Cations, divalent and EcoR
  V endonuclease
  activity, 196
Cau endonucleases, 12
Ccr endonucleases, 12
Cdi27I, 12
CentiMorgan, 279
CflI, 12
CfoI, 13
Cfr endonucleases, 12-13
CfuI, 12
ChiI, 13
Chromatography, of EcoR V
  purification, 191-192
Chromosome complexity
  ladder, 259
Chu endonucleases, 13
ClaI, 11
Cleavage
  by EcoR V endonuclease,
    196-198
  mechanism at site by
    EcoR I endonuclease,
    113-116
  preferential by EcoR I

endonucleases, 111
processive action by
   EcoR I
   endonucleases, 109-
   113
Cleavage frequency
   prediction
   effect of CpG
      frequency in
      mammals, 278
   effect of
      trinucleotide
      usage on, 273
   of restriction
      endonucleases,
      dependance on base
      composition, 270
Cleavage rates, EcoR I
   endonucleases, 114-
   115
Cleavage sequence pairs,
   compatible for
   restriction
   endonucleases, 74-87
Cleavage sites
   for EcoR V restriction
      endonuclease, 186-
      187

generation for
   methylation
   dependent
   endonucleases, 262-
   263
for restriction
   endonucleases, 2-34,
   52-73
Cleavage specificity, of
   BamH I methylases, 267-
   268
Clm endonucleases, 11
Cloning, of pPvu1, 229-
   230
Cloning, of Pst I genes in
   yeast, 219-224
ClfI, 11
CluI, 11
Cotranscripton, of Pst I
   genes, 218
CpaI, 13
CpeI, 13
CpfI, 13
Cross-protection, by
   sequence-specific
   methylases, 258-262
CscI, 11
CueI, 13

CuiI, 12
CunI, 12
Cytosine methylation, PvuII endonuclease activity and, 233

dcm methylase, 239
Dde endonucleases, 13
DdsI, 13
Dimers
  BamH I endonuclease, 153
  EcoR V, 193
Dimethyl sulfate, interaction with left-handed Z-DNA, 252
Dimethylsulfoxide, effect on BamH I endonuclease, 175-176
Dinucleotides
  non-random DNA sequences, 268, 270-271
  usage, effect in Saccharomyces cerevisiae DNA from GENBANK, 271
Dissociation rates, for EcoR I complexes, 107-108
DpnI, 13
DNA
  cleavage. See Cleavage
  hemimethylated
    activity with BamH I endonuclease, 159, 160
    BamH I methylation rates and, 159, 160
  left-handed Z. See Z-DNA
  methylation, sequence-specific, blocking of methylation by, 263-268
  methylation affecting gene expression in E.coli, 236, 238
  non-random sequences, 268-278
  preferential cleavage by EcoR I endonucleases, 108-109
  purity, restriction endonuclease

activity and, 93
specificity and relaxed
specificity, of EcoR
V endonuclease, 200–
204
structure
within EcoRI-DNA
complex, 121–123
in left-handed Z
helix, 248, 249
restriction
endonuclease
activity and, 94
in right-handed B
helix, 248, 249
SV40, resistance to BamH
I endonuclease as
function of methyl
group content, 159,
161
DNA binding
by BamH I methylase, 159–
161
non-specific
electrostatic
component, 165
inhibition of BamH I
endonuclease and

methylase, 168–
169
non-specific flanking,
facilitated
diffusion and, 162–
163
DNA binding proteins,
inhibition of EcoR I
endonuclease cleavage,
113
DNA-protein interactions,
discrimination between
purines and
pyrimidines, 204
DNAse I footprinting of
methylase promoter, in
Pst I restriction-
modification system,
216–217
DNA sequences
for EcoR V restriction-
modification system,
187–189
intragenomic variation,
277
nonspecific, EcoR I
endonuclease
affinity for, 104,

115
prediction of rare
cutters and, 277–
278
for Pst I restriction-
modification system,
210
recognized by EcoR V*,
202–204
specificity,
perturbation by
organic solvents
with BamH I
methylase, 180–181
Dpn endonucleases, 13
Dra endonucleases, 13
Dsa endonucleases, 13

Eae endonucleases, 14
EagI, 14
Eca endonucleases, 14
EccI, 14
Ecl endonucleases, 14
Eco endonucleases, 14–16
EcoR endonuclease
  cleavage sites, 14
  microorganism, 14
  sequences, 14, 104
EcoRI-DNA complex

catalytic clefts, 124–
125
protein interactions
with DNA backbone,
125–127
structural
characteristics, 136–
138
EcoRI endonuclease, 104
base pair recognized by
single interactions,
143–144
base pair specificity
sites, 142
binding to tandem sites,
111
cleavage mechanism at
site, 113–116
combinations giving
unambiguous base
recognition, 145
complex with DNA,
structural
characteristics, 121–
127
conformational change
and specificity, 133–
135

disadvantage in study of
DNA-protein
interactions, 186
electrostatic
interactions and
recognition, 132-
133
equilibrium binding
constants, 105-106
inhibition of cleavage
by DNA binding
proteins, 113
mechanisms of site
location, 106-113
dissociation rates,
107-108
preferential cleavage
of longer DNA
molecules, 108-
109
processive action and,
109-113
primary activity,
mechanism for, 131-
132
processivity, 109-113
topological barriers
to, 111-113

protein interactions
with substrate, 198-
199
reaction profile vs.
EcoR V, 198
recognition mechanism,
127-135
recognition site vs.
EcoR V canonical
site, 204
secondary activity, 175
source, 14
specificity and hydrogen
bonds, 127, 130-131
specificity of binding
and cleavage, 104-
106
structure, 120-145
EcoR 1 linearized pBR 322
DNA, 163
EcoRI methylase, effect of
glycerol on
specificity, 180
EcoRI restriction and
modification system,
104
EcoR V endonuclease
applications in

construction of
recombinant DNA, 186-
187
crystals, 199
DNA cleavage, 196-198
DNA cleavage site, 186-
187
future prospects, 204-
205
gel filtration of, 194-
196
homology with methylase,
187-189
molecular genetics, 187-
189
over-production and
protein
purification, 189-
192
primary and secondary
sites, 202
protein structure and
enzyme mechanism,
192-199
purification, 193
reaction profiles, 197-
198
sedimentation velocity,
194-196
specificity and relaxed
specificity, 200-
204
EcoR V methylase, over-
production and protein
purification, 189-190
EcoR V restriction-
modification system,
molecular genetics of,
187-189
Edman degradation, 214
Effective recognition
sequence size (ERSS)
of BamH I endonuclease,
270
of BamH I methylases,
267-268
for methylases, 264-265
of SMA I endonuclease,
270
of trinucleotides, 273
using
methylase/endonuclease
crossprotection,
260-261
Endonucleases,
restriction. See also

specific
endonucleases
activity, factors
    influencing, 92-100
buffer systems, 87-89
cleavage, ligation and
    sensitivity, 52-100
cleavage sequence pairs
    compatible, 74-87
cleavage sites, 52-73
dependance of cleavage
    frequency on base
    composition, 270
interaction with left-
    handed Z-DNA, 251-
    252
methylation dependent,
    generation of
    cleavage sites for,
    262-263
naming of, 2
recognition sequences,
    52-73
type II
  applications in
    dissection of DNA,
    186
  sequence specificity,

148
study of interactions
    of proteins with
    specific DNA
    sequences, 186
Enzyme stability, for
    restriction
    endonucleases, 89-91
ErpI, 14
ERSS. See Effective
    recognition sequence
    size
Escherichia coli, methyl
    cytosine restriction
    in, 234-243
EspI, 16

Facilitated diffusion,
    Kinetic evidence for
    BamHI enzyme fragments
    of pRB322 DNA, 164-
    165
EbaI, 16
EbrI, 16
Edi endonucleases, 17
Ein endonucleases, 16-17
Flanking sequences, of
    BamH I endonucleases,
    158

Fluorescamine,
    modification of EcoR V
    endonuclease, 198-199,
    200
Enu endonucleases, 17
EokI, 17
Esp endonucleases, 16
EspMI, 17
EsuI, 17

GalI, 17
Gce endonucleases, 17
Gdi endonucleases, 17
GdoI, 17
Gel filtration, of EcoR V
    endonuclease, 194-196
GENBANK, 271
Gene mapping
  at megabase level, 279,
    280
  of Pst I genes, 210
Genetics, molecular, of
    EcoR V restriction-
    modification system,
    187-189
GglI, 18
GinI, 18
Glycerol
  effect on BamH I
    endonuclease, 175-
    176
  effect on EcoR I
    methylase
    specificity, 180
  hypermethylation of BamH
    I methylase and, 180-
    181
GoxI, 18
GsbI, 18
GsuI, 18

HacI, 20
Hae endonucleases, 18
HagI, 20
Hap endonucleases, 18
Heat sensitivity, for
    restriction
    endonucleases, 89-91
Hemimethylation, 266
HgaI, 18
Hgi endonucleases, 20-21
Hha endonucleases, 18,
    251
HhaI methylase, 251
HhgI, 18
Hin endonucleases, 18-19
Hpa endonucleases, 20
Hpa II, methylation, 266

HphI, 20
HsuI, 20
HU proteins, inhibition of
    EcoR I endonuclease
    cleavage, 113
Hydrogen bonds
  EcoRI specificity and,
      127, 130-131
  sequence specific
      protein-DNA
      interactions and,
      172
  specific binding of
      protein to nucleic
      acid residue and,
      178-179
Hydrophobicity, of BamH I
    endonuclease, 154
Hydrophobic residues, 213
5-Hydroxmethyl cytosine,
    234
Hypermethylation, of BamH
    I methylase in
    presence of glycerol,
    180-181

Intersegment transfer, by
    EcoR I endonucleases,
    107

Intradomain transfer, by
    EcoR I endonucleases,
    107
Isoschizomers, of Nru I,
    265
KaeE37I, 21
Kinetics
  of BamH I endonuclease
      with linear DNA
      substrates, 162-
      169
  of BamH I methylase with
      linear DNA
      substrates, 162-169
  hyperbolic, of BamH I
      methylase with form
      I SV40 and pBR322
      DNA, 161-162
Kpn endonucleases, 21

MaE endonucleases, 21
Mapping techniques, 258
MauI, 21
Mbo endonucleases, 22
MbuI, 22
McaI, 22
mcrB locus, 239
Mcr phenotype

characterization of, 239
implications and circumvention of, 239-243
Megabase, level, gene mapping at, 279, 280
Methylase/Dpn I recognition sequences, 264-265
Methylase/M.BamH I/BamH I recognition sequences, 267-268
Methylases
  average fragment size, 264-265
  BamH 1, 148
  EcoR 1, 104
  effective recognition sequence site, 264-265
  interaction with left-handed Z-DNA, 251-252
  PSt I, expression in yeast, 223
  recognition sequences, 264-265
  sequence-specific, cross-protection of restriction endonucleases and, 258-262
Methylase specificity, of BamH I methylases, 267-268
Methylation mechanisms, for BamH I, evaluation with form I SV40 DNA, 159-161
Methyl cytosine restriction, in E. coli, 234-237
Methyl cytosine restriction phenotype. See Mcr phenotype
Methyl group incorporation, BamH I methylase purfication and, 150
O6-Methyl guanine-DNA methyl transferase, interaction with left-handed Z-DNA, 252
MeuI, 21
MflI, 21

MgI endonucleases, 22
Minicells, bacterial,
    synthesis of Pst I
    restriction-
    modification enzymes,
    211-213
MisI, 22
Mja endonucleases, 21
MkiI, 22
MkrI, 21
MlaI, 21
MleI, 21
MltI, 21
MluI, 21
Mni endonucleases, 22
Mnn endonucleases, 22
Mno endonucleases, 22
MnoIII, 23
Mononucleotides, non-
    random DNA sequences,
    268-270
MosI, 23
MphI, 23
MraI, 22
Msi endonucleases, 23
Msp endonucleases, 23
Mst endonucleases, 22
MthI, 21

Mung bean nuclease, 21150,
    215
MuaI, 22
Mui endonucleases, 23

NaeI, 25
NamI, 25
Nan endonucleases, 23
NarI, 25
Nas endonucleases, 25
NbaI, 25
NblI, 25
NbrI, 25
NcaI, 23
NciI, 23
NcoI, 25
NcuI, 23
NdaI, 25
Nde endonucleases, 23
Nfl endonucleases, 23
NflIII, 24
NgbI, 25
Ngo endonucleases, 24
NheI, 24
Nicked reaction
    intermediates,
    dissociation of, 115
Nla endonucleases, 24
Nme endonucleases, 24

NmiI, 25
Nmu endonucleases, 24-25
NocI, 26
NotI, 26
Nou endonucleases, 25
Npo endonucleases, 26
NruI, 26
Nsi endonucleases, 25
NspAI, 26
Nsp endonucleases, 26-27
Nsu endonucleases, 25
Nta endonucleases, 26
Nun endonucleases, 26

Olignonucleotides,
    synthetic, in
    derivation of EcoR V
    endonuclease, 186
Open-reading frames, for
    EcoR V restriction-
    modification system,
    187-189
Oxa endonucleases, 27

Pae endonucleases, 27
PaiI, 27
PalI, 27
PanI, 27
Periplasmic space, 213

PfaI, 27
Pfl endonucleases, 27
Pgl endonucleases, 28
Plasmids
    in derivation of EcoR V
        endonuclease, 186
    pAT15322, EcoR V
        activity and, 196-
        197
    pBR322
        methylation, effects
            on transformation
            of mcr+- strains,
            242-243
        recombinant DNA
            vector, 186-187
    pBR322 DNA, lacking BamH
        1 site, 168
    pBR322L, EcoR I
        endonuclease
        cleavage of, 113-
        114
    pBR32(RI)2
        processive action of
            EcoR 1
            endonuclease and,
            109-111
        structure, 109

pPst201, 214
pPvuI, 233-234
pPvuI, cloning, 229-230
pYE320, cloning of foreign genes into yeast, 223-224
pYE002, transformation of Saccharomyces cerevisiae, 220
pYE004, transformation of Saccharomyces cerevisiae, 220
recombinant, 221
Pma endonucleases, 28
PmiI, 28
PmyI, 27
PouI, 28
PpaI, 28
Ppu endonucleases, 28
Preditor, best, non-random DNA sequences, 271, 275
Pribnow box, 215
Processivity experiments, for EcoR I endonucleases, 116
Promoters
  ACCI, 221-224

E. coli consensus, 188
PL, 189
Proteins
  DNA binding, inhibition by EcoR endonuclease cleavage, 113
  interactions with DNA backbone, 125-127
  interactions with substrate, EcoR 1 endonuclease, 198-199
  of Pst I restriction-modification system, 211-214
  structural organization within EcoRI-DNA complex, 123-124
PspI, 27
Psp6II, 28
Pss endonucleases, 28
PstI, 27
Pst I gene, cloning in yeast, 219-224
Pst 1 genes, control of expression, 218-219
Pst I restriction-modification system

determination of
  nucleotide sequence,
  210
DNAse I footprinting of
  methylase promoter,
  216-217
organization and control
  of expression, 210-
  224
proteins, 211-214
transcription of
  initiation sites,
  215
Purification
  of BamH I endonuclease,
  148-150
  of BamH I methylase, 150-
  152
Pvu endonucleases, 27
Pvu II methylase, 228
  characterization of, 232-
  233
Pvu II restriction-
  modification system,
  228, 228-243
  characterization of Pvu
  II gene products,
  232-233

cloning of pPvu1, 229-
  230
methyl cytosine
  restriction in E.
  coli, 234-243
relative positions of
  Pvu II genes, 230-
  232
RMS2s, 229
Pvu IIRMS2 endonuclease,
  228

Recognition sequences
  methylase/M.BamH I/BamH
  I, 267-268
  for methylases, 2-34,
  264-265
  for restriction
  endonucleases, 2-34,
  52-73
Recombinant plasmids, 221
Recombination, site
  specific, mediation by
  EcoR I endonuclease,
  219
recl protein, interaction
  with left-handed Z-
  DNA, 252
Repressor, 189

RheI, 28
Rhp endonucleases, 28
RhsI, 29
RleI, 28
RluI, 28
RmeI, 28
RNA polymerase
  formation of complex
    with methylase
    promoter, 217, 219
  interaction with left-
    handed Z-DNA, 252
RrbI, 29
Rrh endonucleases, 28
RroI, 28
RsaI, 29
Rsh endonucleases, 29
RspI, 29
Rsr endonucleases, 29

SaaI, 31
SabI, 31
Sac endonucleases, 30-31
Sal endonucleases, 31
SaOI, 31
Sau endonucleases, 30, 31
SB, 29
Sbo13, 29
SboI, 31

ScaI, 31
SciI, 32
SciNI, 30
ScoI, 31
ScrFL, 30
ScuI, 31
SduI, 30
SdyI, 30
Sec endonucleases, 33
Sedimentation velocity, of
  EcoR V, 194-196
Sex endonucleases, 31
Sfa endonucleases, 30
SfiI, 31
SflI, 32
SfnI, 29
SfoI, 29
SfrI, 31
SgaI, 31
SgoI, 31
Sgr endonucleases, 32
Shine-Dalgarno sequence,
  215
Shy endonucleases, 32
Signal peptide, 214
SinI, 29
SinM endonucleases, 30
Ska endonucleases, 32

SlaI, 32
SluI, 32
SmaI, 29, 270
SnaBI, 29
SnaI, 30
SnoI, 32
Sod endonucleases, 32
SP, 29
Spa endonucleases, 32
SpaI, 31
SpeI, 30
SphI, 32
Spl endonucleases, 30
SQ, 29
SsaI, 30
Sso endonucleases, 29
SspI, 30
SspXI, 29
Sst endonucleases, 32
StuI, 32
StxI, 29
SuaI, 33
SuiI, 33

TaqI endonucleases, 33
TecI, 33
Temperature, assay, 89-91
Tetramers, of BamH I
    endonuclease, 153

TflI, 33
TglI, 33
ThaI, 33
Transcription
    for EcoR V restriction-
        modification system,
        188-189
    of initiation sites in
        Pst 1 restriction-
        modification system,
        215
Transfer mechanisms, 107,
    109
Transformation, of S.
    cerevisiae by plasmid
    pYE004 and pYE002,
    220
Transformation
    efficiencies, of LL20
    with plasmid DNA, 220
Translation, for EcoR V
    restriction-
    modification system,
    188
Trinucleotides
    non-random DNA
        sequences, 271-274
    in sequenced DNA from

variety of species, 275-276
ItaI, 33
Ith endonucleases, 33
IthHB8I, 33
ItnI, 33
ItrI, 33

Velocity studies, with BamH I methylase, 161-162
UhaI, 34

XamI, 34
XbaI, 34
XciI, 34
XcyI, 34
Xho endonucleases, 34
Xma endonucleases, 34
XmnI, 34
XniI, 34
Xor endonucleases, 34
XpaI, 34
XphI, 34

Yeast, cloning of Pst I gene, 219-224

ZanI, 34
Z-DNA

enzymatic probes for, 248-253
interaction
  with BAL31 nuclease, 250-251
  with dimethyl sulfate, 252
  with mung bean nuclease, 250
  with P1 nuclease from *Penicillium citrinum*, 251
  with rec1 protein from *Ustilago*, 252
  with repair enzymes, 252
  with restriction endonucleases and methylases, 251-252
  with RNA polymerase, 252
  with S1 nuclease from *Aspergillus oryzae*, 248, 250
structure, 248, 249